THE THEORY OF BEST APPROXIMATION AND FUNCTIONAL ANALYSIS

IVAN SINGER
Institute of Mathematics
Bucharest

SOCIETY for INDUSTRIAL and APPLIED MATHEMATICS
PHILADELPHIA, PENNSYLVANIA 19103

Printed for the Society for Industrial and Applied Mathematics by
J. W. Arrowsmith Ltd., Bristol 3, England

Contents

Preface

In this monograph we present some results and problems in the modern theory of best approximation, i.e., in which the methods of functional analysis are applied in a consequent manner. This modern theory constitutes both a unified foundation for the classical theory of best approximation (which treats the problems with the methods of the theory of functions) and a powerful tool for obtaining new results. Within the general framework of normed linear spaces the problem of best approximation amounts to the minimization of a distance, which permits us to use geometric intuition (but rigorous analytic proofs), and the connections of the phenomena become clearer and the arguments simpler than those of the classical theory of best approximation in the various particular concrete function spaces. We hope that this has been proved convincingly enough in the monograph [168] (which was the first of this kind in the literature) and in the lecture notes [175], and will be proved again in the present monograph (see, for example, § 1, the remarks made after Theorem 1.8, or § 3, the remark to Theorem 3.5).

Naturally, the interaction between the theory of best approximation and functional analysis works also in the other direction, for example, problems of best approximation in normed linear spaces have led to the discovery of the theorem on extremal extension of extremal functionals, of the concrete representations for the extremal points of the unit cell in certain conjugate spaces, of new results on semi-continuity of set-valued mappings, etc. However, we shall not consider this side of their interaction in the present monograph.

Since June 1966, when the Romanian version of the monograph [168] went to print, the theory of best approximation in normed linear spaces has developed rapidly. However, except for the expository paper, covering the period up to 1967, of A. L. Garkavi [67], the only attempt for a comprehensive survey material was made in [175]. The latter constitutes also the basis for the present monograph, which, though self-contained, may be regarded as an up-to-date complement to [168] and [175] ; we have essentially conserved the structure of [175], but improved the parts which overlap with [175] and added much new material, part of which appeared after [175] went to print (the bibliography of [175] contained only 99 items).

We note also the appearance, since 1970, of the lecture notes of A. L. Brown [34], P. D. Morris [132] and R. B. Holmes [82] and of the book of P. J. Laurent [118], which treat some topics in the theory of best approximation in normed linear spaces and optimization in locally convex spaces [118], [82] ; let us also mention the lecture notes of F. Deutsch (*Theory of Approximation in Normed Linear Spaces*,

Fall 1972), which we received when the present work was completed. We take the liberty of suggesting that it would be more appropriate to replace "Approximation" by "Best Approximation" in the titles of all these works with the exception of [82]; indeed, since these works (except for part of [118]) are actually concerned with the theory of best approximation, which is only a part of approximation theory, their present titles might mislead the reader to expect more topics to be covered. Let us note that the use of the word "best" in the titles of [168], [67], [175], [82] and of the present monograph, which serves also for the delimitation of their scope (namely, to emphasize that some other important parts of approximation theory are not considered), is not a lack of modesty, but a universally accepted classical term for approximation problems related to nearest points; this is the term which is used also in the texts of the abovementioned works of Brown, Morris, Laurent and Deutsch.

In order to be able to enter more deeply into the problems, without increasing the size of the present monograph, we have restricted ourselves to consider only five basic topics in the theory of best approximation, shown in the titles of the five sections. Thus, various problems of best approximation treated in [168] (for example, the distance functionals, metric projections onto sequences of subspaces, deviations of sets from linear subspaces, finite-dimensional diameters and secants, Chebyshev centers, etc.) and not treated in [168] (for example, methods of computation of elements of best approximation, applications to extremal problems of the theory of analytic functions, etc.) have been deliberately omitted. On the other hand, [168] contained only a few isolated results on metric projections (for example, we did not include in [168] some of the results of [166], [122], [32]), since their theory was only beginning to develop at that time, but in [175] and in the present monograph the largest section is devoted to metric projections (see § 4). Also, [168] contained only a short appendix on best approximation by elements of nonlinear sets, since this topic was beyond the scope of [168], but in the present monograph these problems, although treated briefly, occupy the second largest section (see § 5).

Furthermore, in order to limit the size of the present monograph, we have included here only a few short proofs (some of them being very simple and some, which are less elementary, being particularly interesting), but for all results we give references. Also, we do not present all known applications of the general theory in concrete spaces, but only some examples of some more important ones. Finally, we tried to reduce the overlapping with [168] to the very minimum necessary for self-containedness and thus we often refer to [168] for complementary results and bibliography. In the bibliography of the present monograph we wanted to emphasize those works which have appeared after [168] went to print and some earlier papers which have been omitted from [168]; therefore, sometimes we give here some results (and their authors' names) only with reference to the bibliography of [168]. In exchange for these limitations, we have tried to give the up-to-date stand and literature of the problems treated herein.

We hope that the present monograph will be useful for a large circle of readers, including those who are not specialists in these problems and those specialists who work in the theory of best approximation with the classical function-theoretic methods or using the methods of functional analysis. Also, for specialists in functional analysis, in particular, in the geometry of normed linear spaces, this monograph may offer a new field of applications.

The reader is assumed to know some elements of functional analysis and integration theory, but we recall (giving also a reference to a treatise), whenever necessary, the notions and results which we use.

We acknowledge with pleasure that we benefited from attending the seminar lectures of Dr. G. Godini (at the Institute of Mathematics of the Academy, Bucharest, in 1971–1973) and from our visits at the universities of Bonn (1971, 1972) and Grenoble (1972), where we had useful discussions with Dr. W. Pollul and Professors J. Blatter and P.-J. Laurent on some problems treated in this monograph. Finally, we wish to express our thanks to Professor R. S. Varga for the invitation to write this monograph and to present it at the National Science Foundation Regional Conference on Theory of Best Approximation and Functional Analysis at Kent State University, June 11–June 15.

IVAN SINGER

Bucharest

June 1, 1973

The Theory of Best Approximation and Functional Analysis

Ivan Singer

1. Characterizations of elements of best approximation.

(a) Throughout the sequel, without any special mention, we shall denote by ρ the distance in a metric space E and, in particular, if E is a normed linear space, ρ will denote the distance in E induced by the norm, i.e.,

$$\rho(x, y) = \|x - y\|, \qquad x, y \in E. \tag{1.1}$$

DEFINITION 1.1. Let E be a metric space, G a set in E and $x \in E$. An element $g_0 \in G$ is called an *element of best approximation* of x (by the elements of the set G) if we have

$$\rho(x, g_0) = \inf_{g \in G} \rho(x, g), \tag{1.2}$$

i.e., if g_0 is "nearest" to x among the elements of G; we shall denote by $\mathscr{P}_G(x)$ the set of all such elements g_0, i.e.,

$$\mathscr{P}_G(x) = \{g_0 \in G | \rho(x, g_0) = \inf_{g \in G} \rho(x, g)\}. \tag{1.3}$$

It is natural to consider first the problem of characterization of elements of best approximation, i.e., the problem of finding necessary and sufficient conditions in order that $g_0 \in \mathscr{P}_G(x)$, since these results will be applied to solve the other problems on best approximation (for example, those of existence and uniqueness of elements of best approximation, etc.). Also, the characterization theorems in concrete spaces (see, for example, the "alternation theorem" 1.13 below) are convenient tools for verifying whether or not a given g_0 satisfies $g_0 \in \mathscr{P}_G(x)$, since they are easier to use than (1.2).

Since we have obviously

$$\mathscr{P}_G(x) = \begin{cases} x & \text{for } x \in G, \\ \varnothing & \text{for } x \in \overline{G} \setminus G, \end{cases} \tag{1.4}$$

it will be sufficient to characterize the element of best approximation of the elements $x \in E \setminus \overline{G}$. In order to exclude the case when such elements do not exist, in the sequel we shall assume, without any special mention, that $\overline{G} \neq E$.

Unless otherwise stated, the field of scalars for all (general or concrete) normed linear spaces considered in the sequel can be either the field of complex numbers or the field of real numbers.

(b) The first main theorem of characterization of elements of best approximation by elements of linear subspaces in normed linear spaces is the following (see [168, p. 18]):

THEOREM 1.1. *Let E be a normed linear space, G a linear subspace of E, $x \in E \setminus \overline{G}$ and $g_0 \in G$. We have $g_0 \in \mathscr{P}_G(x)$ if and only if there exists an $f \in E^*$ such that*

$$\|f\| = 1, \tag{1.5}$$

$$f(g) = 0, \qquad g \in G, \tag{1.6}$$

$$f(x - g_0) = \|x - g_0\|. \tag{1.7}$$

We recall that E^* denotes the conjugate space of E, i.e., the space of all continuous linear functionals on E, endowed with the usual vector operations and with the norm $\|f\| = \sup_{\substack{x \in E \\ \|x\| \leqq 1}} |f(x)|$.

To prove Theorem 1.1, assume that $g_0 \in \mathscr{P}_G(x)$. Then, since $x \in E \setminus \overline{G}$, we have $\rho(x, G) = \|x - g_0\| > 0$ and hence, by a corollary of the Hahn–Banach theorem (see, for example, [55, p. 64, Lemma 12]), there exists an $f_0 \in E^*$ such that $\|f_0\| = 1/\|x - g_0\|$, $f_0(g) = 0$ $(g \in G)$, and $f_0(x) = 1$. Then the functional $f = \|x - g_0\| f_0 \in E^*$ satisfies (1.5)–(1.7). Conversely, if there is an $f \in E^*$ satisfying (1.5)–(1.7), then for any $g \in G$ we have

$$\|x - g_0\| = |f(x - g_0)| = |f(x - g)| \leqq \|f\| \, \|x - g\| = \|x - g\|$$

and hence $g_0 \in \mathscr{P}_G(x)$, which completes the proof.

It is easy to see that Theorem 1.1 admits the following *geometrical interpretation*: *We have $g_0 \in \mathscr{P}_G(x)$ if and only if there exists a closed hyperplane G' in E* (i.e., a closed linear subspace G' such that $\dim E/G' = 1$) *containing G, which supports the cell* $S(x, \|x - g_0\|) = \{y \in E \mid \|y - g_0\| \leqq \|x - g_0\|\}$ (i.e., $\rho(G', S(x, \|x - g_0\|)) = 0$ and $G' \cap \operatorname{Int} S(x, \|x - g_0\|) = \emptyset$).

Any functional $f \in E^*$ satisfying (1.5) and (1.7) is called a "maximal functional" of the element $x - g_0$ (because we have $\|x - g_0\| = \sup_{\substack{h \in E^* \\ \|h\| = 1}} |h(x - g_0)|$). The usefulness of Theorem 1.1 for applications in various concrete normed linear spaces is due to the fact that for these spaces the general form of maximal functionals of the elements of the space is well known and simple (see, for example, [154], [198]).

We recall that an element x of a closed convex set A in a topological linear space L is called an *extremal point* of A if the relations $y, z \in A$, $0 < \lambda < 1$, $x = \lambda y + (1 - \lambda)z$ imply $y = z = x$. The second main theorem of characterization of elements of best approximation by elements of linear subspaces in normed linear spaces is the following (see [168, p. 62]).

THEOREM 1.2. *Let E be a normed linear space, G a linear subspace of E, $x \in E \setminus \overline{G}$ and $g_0 \in G$. We have $g_0 \in \mathscr{P}_G(x)$ if and only if for every $g \in G$ there exists an extremal point f^g of the unit cell $S_{E^*} = \{f \in E^* \mid \|f\| \leqq 1\}$ such that*

$$\operatorname{Re} f^g(g - g_0) \leqq 0, \tag{1.8}$$

$$f^g(x - g_0) = \|x - g_0\|. \tag{1.9}$$

For a geometrical interpretation of Theorem 1.2 see [168, p. 75]. Theorem 1.2 is also convenient for applications in the usual concrete spaces because for these spaces the general form of the extremal points of S_{E^*} is well known and simple.

It is easy to see that the sufficiency part of Theorem 1.2 remains valid for an arbitrary set G in E. Indeed, if the condition is satisfied, then for every $g \in G$ we have

$$\begin{aligned}\|x - g_0\| &= \operatorname{Re} f^g(x - g_0) \leqq \operatorname{Re} f^g(x - g) \leqq |f^g(x - g)| \\ &\leqq \|f^g\| \, \|x - g\| = \|x - g\|,\end{aligned}$$

whence $g_0 \in \mathscr{P}_G(x)$, which proves the assertion. The problem of characterizing those sets $G \subset E$ for which the condition in Theorem 1.2 is also necessary, is important in nonlinear approximation (see § 5).

Since best approximation amounts, by definition, to the minimization of the convex functional $\varphi = \varphi_x$, on the linear subspace G of a normed linear space E, where

$$\varphi(y) = \|x - y\|, \qquad y \in E, \tag{1.10}$$

there naturally arises the problem of obtaining characterizations of the elements of best approximation $g_0 \in \mathscr{P}_G(x)$ with the aid of differential calculus. The main difficulty is that in general the norm in E is not necessarily Gâteaux differentiable at each nonzero point of E. Nevertheless, it is known (see, for example, [55, p. 445, Lemma 1]) that the limits

$$\tau(x, y) = \lim_{t \to 0^+} \frac{\|x + ty\| - \|x\|}{t}, \qquad x, y \in E, \tag{1.11}$$

always exist and one can use them to give the following characterizations of elements of best approximation (see [168, pp. 88–90]).

THEOREM 1.3. *Let E be a normed linear space, G a linear subspace of E, $x \in E \setminus \overline{G}$ and $g_0 \in G$. We have $g_0 \in \mathscr{P}_G(x)$ if and only if*

$$\tau(x - g_0, g) \geqq 0, \qquad g \in G. \tag{1.12}$$

If the norm in E is Gâteaux differentiable at $x - g_0$, this condition is equivalent to the following:

$$\tau(x - g_0, g) = 0, \qquad g \in G. \tag{1.13}$$

Let us observe that Theorem 1.1 can be also expressed as follows: $g_0 \in \mathscr{P}_G(x)$ if and only if $0 \in \{f|_G \in G^* \mid f \in S_{E^*}, f(x - g_0) = \|x - g_0\|\}$. We have also the following characterization theorem [171].

THEOREM 1.4. *Let E be a normed linear space, G a linear subspace of E, $x \in E \setminus \overline{G}$ and $g_0 \in G$. We have $g_0 \in \mathscr{P}_G(x)$ if and only if 0 belongs to the $\sigma(G^*, G)$-closure of the* *convex hull of the set*

$$\Lambda = \{\overline{f(x-g_0)f|_G \in G^* | f \in \mathscr{E}(S_{E^*}), |f(x-g_0)| = \|x-g_0\|}\}. \tag{1.14}$$

Theorems 1.1–1.4 above can be also expressed, more concisely, as formulae for $\mathscr{P}_G(x)$, namely as

$$\mathscr{P}_G(x) = \{g_0 \in G | \, \|x-g_0\| = \max_{\substack{f \in G^\perp \\ \|f\|=1}} |f(x-g_0)|\}, \tag{1.15}$$

$$\mathscr{P}_G(x) = \{g_0 \in G | \min_{\substack{f \in \mathscr{E}(S_{E^*}) \\ f(x-g_0)=\|x-g_0\|}} \operatorname{Re} f(g-g_0) \leqq 0 \, (g \in G)\}, \tag{1.16}$$

$$\mathscr{P}_G(x) = \{g_0 \in G | \tau(x-g_0, g) \geqq 0 \, (g \in G)\}, \tag{1.17}$$

$$\mathscr{P}_G(x) = \{g_0 \in G | 0 \in \widetilde{\operatorname{co}} \, \Lambda\}, \tag{1.18}$$

respectively, where we use the notation

$$G^\perp = \{f \in E^* | f(g) = 0 \, (g \in G)\}, \tag{1.19}$$

$\mathscr{E}(S_{E^*})$ for the set of all extremal points of S_{E^*}, and $\widetilde{\operatorname{co}}$ for the weak* closed convex hull, and where max, min stand for such sup, inf which are attained. Note also the obvious relations

$$\mathscr{P}_G(x) = G \cap S(x, \rho(x, G)) = G \cap \operatorname{Fr} S(x, \rho(x, G)), \tag{1.20}$$

which are useful for geometrical interpretations. Finally, let us give a characterization theorem (which we shall express concisely as formulae for $\mathscr{P}_\Gamma(f)$ and $\mathscr{P}_G(x)$), which will be useful in the study of existence and uniqueness problems (§§ 2–3) and of the set-valued metric projection operators (§ 4).

THEOREM 1.5. *Let E be a normed linear space.*

(i) *If Γ is a $\sigma(E^*, E)$-closed linear subspace of E^* and $f \in E^* \setminus \Gamma$, then*

$$\mathscr{P}_\Gamma(f) = f - \{h \in E^* | \, h|_{\Gamma_\perp} = f|_{\Gamma_\perp}, \|h\| = \|f|_{\Gamma_\perp}\|\}, \tag{1.21}$$

where

$$\Gamma_\perp = \{x \in E | \gamma(x) = 0 \, (\gamma \in \Gamma)\}. \tag{1.22}$$

(ii) *If G is a closed linear subspace of E and $x \in E \setminus G$, then*

$$\mathscr{P}_G(x) = x - \{y \in E | \, \kappa(y)|_{G^\perp} = \kappa(x)|_{G^\perp}, \|y\| = \|\kappa(x)|_{G^\perp}\|\}, \tag{1.23}$$

where κ denotes the canonical isometrical embedding of E into E^{**}, that is,

$$(\kappa(x))(f) = f(x), \qquad x \in E, f \in E^*. \tag{1.24}$$

Part (i) has been given in [174, formula (25)']. Here is a slightly more direct proof, using an argument of R. R. Phelps [145] : If $\gamma_0 \in \mathscr{P}_\Gamma(f)$, then for $h = f - \gamma_0$, the element $f - h$ is in the right-hand side of (1.21), since by the Hahn-Banach theorem

$$\|f|_{\Gamma_\perp}\| = \min_{\gamma \in (\Gamma_\perp)^\perp} \|f - \gamma\| = \min_{\gamma \in \Gamma} \|f - \gamma\|$$

(we have $(\Gamma_\perp)^\perp = \Gamma$, since Γ is $\sigma(E^*, E)$-closed). Conversely, if $\gamma_0 = f - h$ is in the right-hand side of (1.21), then $\gamma_0 = f - h \in (\Gamma_\perp)^\perp = \Gamma$ and, by the preceding formula for $\|f|_{\Gamma_\perp}\|$, we have $\gamma_0 \in \mathscr{P}_\Gamma(f)$. To see part (ii), let $g_0 \in \mathscr{P}_G(x)$. Then for $y = x - g_0$, the element $x - y$ is in the right-hand side of (1.23), by virtue of (1.15). Conversely, if $g_0 = x - y$ is in the right-hand side of (1.23), then, since G is closed, we have $g_0 = x - y \in (G^\perp)_\perp = \bar{G} = G$ and hence, again by (1.15), $g_0 \in \mathscr{P}_G(x)$, which proves part (ii). Note that Theorem 1.5 implies that

$$\kappa(\mathscr{P}_G(x)) \subset \kappa(x) - \{\Phi \in E^{**} |\ \Phi|_{G^\perp} = \kappa(x)|_{G^\perp}, \|\Phi\| = \|\kappa(x)|_{G^\perp}\|\} = \mathscr{P}_{G^{\perp\perp}}(\kappa(x)). \tag{1.25}$$

For a characterization of elements of best approximation in terms of fixed points of a set-valued mapping, see § 5, Proposition 5.1.

(c) Let us give now examples of applications of some of the above theorems in concrete spaces.

We shall use the word "compact" in the sense of N. Bourbaki, i.e., bicompact Hausdorff. For a compact space Q we shall denote by $C(Q)$, respectively $C_R(Q)$, the space of all complex or real, respectively of all real, continuous functions on Q, endowed with the usual vector operations and with the norm $\|x\| = \max_{q \in Q}|x(q)|$. Using the general form of maximal functionals of the elements of $C_R(Q)$, from Theorem 1.1 we obtain (see [168, p. 33])

THEOREM 1.6. *Let $E = C_R(Q)$ (Q compact), G a linear subspace of E, $x \in E \setminus \bar{G}$ and $g_0 \in G$. We have $g_0 \in \mathscr{P}_G(x)$ if and only if there exist two disjoint closed subsets $Y_{g_0}^+$, $Y_{g_0}^-$ of Q and a Radon measure μ on Q, such that*

$$|\mu|(Q) = 1, \tag{1.26}$$

$$\int_Q g(q)\, d\mu(q) = 0, \qquad g \in G, \tag{1.27}$$

$$\mu \geqq 0 \text{ on } Y_{g_0}^+, \quad \mu \leqq 0 \text{ on } Y_{g_0}^- \text{ and } Y_{g_0}^+ \cup Y_{g_0}^- \supset S(\mu), \tag{1.28}$$

$$x(q) - g_0(q) = \begin{cases} \|x - g_0\| & \text{for } q \in Y_{g_0}^+, \\ -\|x - g_0\| & \text{for } q \in Y_{g_0}^-, \end{cases} \tag{1.29}$$

where $S(\mu)$ denotes the carrier of the measure μ.

One can also give a characterization theorem in the spaces $E = C(Q)$ (see 168, p. 29]). Theorem 1.6, which appeared in [165], has constituted the first theorem of characterization in $E = C_R(Q)$ (even in $E = C_R([a, b])$) of elements of best approximation by elements of linear subspaces *G of arbitrary dimension.* From Theorem 1.6 one obtains easily, for example, the following result of E. W. Cheney and A. A. Goldstein (see [168, p. 44]) : *Let L be a real topological linear space, Q a compact subset of L, $E = C_R(Q)$, $G = L^*|_Q$, $x \in E \setminus G$ and $g_0 = \Phi_0|_Q \in G$, where $\Phi_0 \in L^*$. We have $g_0 \in \mathscr{P}_G(x)$ if and only if $0 \in$ co B, where*

$$B = \{q \in Q | x(q) - g_0(q) = \|x - g_0\|\} \cup -\{q \in Q | x(q) - g_0(q) = -\|x - g_0\|\},$$

and where co *stands stands for "convex hull".*

For a positive measure space (T, ν) (we shall not specify the σ-field of subsets of T on which the measure ν is defined; this will cause no confusion) and for $1 \leqq p < \infty$ (respectively, $p = \infty$) we shall denote by $L^p(T, \nu)$ the space of all equivalence classes of functions with ν-integrable pth power (respectively of ν-measurable and ν-essentially bounded functions on T), endowed with the usual vector operations and with the norm $\|x\| = (\int_T |x(t)|^p \, d\nu(t))^{1/p}$ (respectively, $\|x\| = \text{ess sup}_{t \in T} |x(t)|$); for simplicity, we use here the same notation for a function and for its equivalence class in L^p. Again the subscript R will mean, both here and for the spaces occurring in the sequel, that we restrict ourselves to real scalars. For a function x' on T we shall use the notation

$$Z(x') = \{t \in T | x'(t) = 0\}. \tag{1.30}$$

Using the general form of maximal functionals of the elements of $L^1(T, \nu)$, we obtain from Theorem 1.1 the following theorem (see [168, p. 46]), which was obtained initially by B. R. Kripke and T. J. Rivlin [110] with different (function-theoretic) methods and with the above functional analytic methods in [167].

THEOREM 1.7. *Let $E = L^1(T, \nu)$ (where (T, ν) is a positive measure space), G a linear subspace of E, $x \in E \setminus \bar{G}$ and $g_0 \in G$. We have $g_0 \in \mathscr{P}_G(x)$ if and only if*

$$\left| \int_{T \setminus Z(x - g_0)} g(t) \operatorname{sign} [x(t) - g_0(t)] \, d\nu(t) \right| \leqq \int_{Z(x - g_0)} |g(t)| \, d\nu(t), \qquad g \in G. \tag{1.31}$$

We recall that for a complex number $\alpha \neq 0$, by definition $\operatorname{sign} \alpha = e^{-i \arg \alpha} = \bar{\alpha}/|\alpha|$ and that $\operatorname{sign} 0 = 0$.

For $E = L^p(T, \nu)$ with $1 < p < \infty$ and for an abstract inner product space H we obtain from Theorem 1.1 the following well-known results (see [168, pp. 56–57]).

THEOREM 1.8. (i) *Let $E = L^p(T, \nu)$ (with (T, ν) a positive measure space and $1 < p < \infty$), G a linear subspace of E, $x \in E \setminus \bar{G}$ and $g_0 \in G$. We have $g_0 \in \mathscr{P}_G(x)$ if and only if*

$$\int_T g(t) |x(t) - g_0(t)|^{p-1} \operatorname{sign} [x(t) - g_0(t)] \, d\nu(t) = 0, \qquad g \in G. \tag{1.32}$$

(ii) *Let $E = H$ be an inner product space, G a linear subspace of E, $x \in E \setminus \bar{G}$ and $g_0 \in G$. We have $g_0 \in \mathscr{P}_G(x)$ if and only if*

$$(g, x - g_0) = 0, \qquad g \in G, \tag{1.33}$$

where (x, y) denotes the scalar product in $E = H$.

Some applications of Theorem 1.1 in other concrete spaces are given in [168].

The above results in concrete spaces illustrate the power of the methods of functional analysis in the theory of best approximation. Indeed, although Theorem 1.1 in general normed linear spaces is obtained by a simple application of a corollary of the Hahn-Banach theorem, it gives in various concrete spaces Theorems 1.6–1.8 and other results as particular cases, simply by using the general form of maximal functionals in these spaces. A direct proof of Theorems 1.6–1.8 would require different methods in each of the concrete spaces involved, apparently having no connection with each other; however, they all turn out to be particular cases of

Theorem 1.1, and this unified method of obtaining them is simpler and clearer than the separate proofs in each concrete space. This unified method is carried out in a consequent manner for the whole theory of best approximation (e.g., for problems of uniqueness, etc.), in the monograph [168]. In the sequel we shall only indicate some examples of applications in concrete spaces (rather than mention all known applications) of some of the general results on best approximation in arbitrary normed linear spaces which we shall give. Let us give now an application of Theorem 1.2.

It is well known (see, for example, [55, p. 441, Lemma 6]) that if $E = C(Q)$, a functional $f \in E^*$ is an extremal point of S_{E^*} if and only if there exist a $q \in Q$ and a scalar α with $|\alpha| = 1$ such that

$$f(y) = \alpha y(q), \qquad y \in E = C(Q). \tag{1.34}$$

From Theorem 1.2, using the general form (1.34) of the extremal points of S_{E^*} for $E = C(Q)$, one obtains immediately the following classical theorem of characterization of elements of best approximation, due to A. N. Kolmogorov (see [168, p. 69]).

THEOREM 1.9. *Let $E = C(Q)$ (Q compact), G a linear subspace of E, $x \in E \setminus \bar{G}$ and $g_0 \in G$. We have $g_0 \in \mathcal{P}_G(x)$ if and only if for every $g \in G$ there exists a $q = q^g \in Q$ such that*

$$\operatorname{Re} \overline{[x(q) - g_0(q)]}g(q) \geqq 0, \tag{1.35}$$

$$|x(q) - g_0(q)| = \|x - g_0\|. \tag{1.36}$$

For applications of Theorem 1.2 in other concrete spaces, see [168, pp. 75–87]. In $E = C(Q)$, using (1.34), one obtains from Theorem 1.4 a theorem of Y. Ikebe [94].

(d) Now we shall consider the important particular case when $\dim G = n < \infty$, i.e., when $G = [x_1, \cdots, x_n]$ = the (closed) linear subspace of E spanned by n linearly independent elements $x_1, \ldots, x_n$. Naturally, the preceding results are also applicable in this particular case; however, by using effectively the assumption $\dim G = n < \infty$, we can obtain additional information. For example, using the classical theorem of Minkowski that the elements of the unit cell

$$S_{E^*} = \{f \in E^* | \, \|f\| \leqq 1\}$$

in a finite-dimensional space E^* can be expressed as finite convex combinations of extremal points of S_{E^*} and that if G is an arbitrary linear subspace of a normed linear space E, then every extremal point of the unit cell S_{G^*} can be extended to an extremal point of S_{E^*} (see, for example, [168, p. 168]), from Theorem 1.1 we obtain (see [168, p. 170]):

THEOREM 1.10. *Let E be a normed linear space, $G = [x_1, \cdots, x_n]$ an n-dimensional linear subspace of E, $x \in E \setminus G$ and $g_0 \in G$. We have $g_0 \in \mathcal{P}_G(x)$ if and only if there exist h extremal points $f_1, \cdots, f_h$ of the unit cell S_{E^*}, where $1 \leqq h \leqq n + 1$ if the scalars are real and $1 \leqq h \leqq 2n + 1$ if the scalars are complex and h numbers*

$\lambda_1, \cdots, \lambda_h > 0$ *with* $\sum_{j=1}^h \lambda_j = 1$, *such that*

$$(1.37) \qquad \sum_{j=1}^{h} \lambda_j f_j(g) = 0, \qquad g \in G,$$

$$(1.38) \qquad \sum_{j=1}^{h} \lambda_j f_j(x - g_0) = \|x - g_0\|.$$

In other words, the additional information to Theorem 1.1 which we obtain for $\dim G = n < \infty$ is that for such subspaces G one can take the functional f of Theorem 1.1 to be a convex combination $f = \sum_{j=1}^h \lambda_j f_j$ of $h \leqq n + 1$ (respectively $h \leqq 2n + 1$) extremal points of the unit cell $S_{E^*} = \{f \in E^* \mid \|f\| \leqq 1\}$.

Using the mapping $\psi \to \{\psi(x_i)\}_1^n$ of G^* onto the n-space, from Theorem 1.4 one obtains [171]:

THEOREM 1.11. *Let E be a normed linear space, $G = [x_1, \cdots, x_n]$ an n-dimensional linear subspace of E, $x \in E \setminus G$ and $g_0 \in G$. We have $g_0 \in \mathscr{P}_G(x)$ if and only if 0 belongs to the convex hull of the following set in the n-dimensional Euclidean space:*

$$(1.39) \quad B_1 = \{\{\overline{f(x - g_0)}f(x_1), \cdots, \overline{f(x - g_0)}f(x_n)\} \mid f \in \mathscr{E}(S_{E^*}), |f(x - g_0)| = \|x - g_0\|\}.$$

(e) Let us give some applications in concrete spaces.

We recall that a system of n functions $x_1, \cdots, x_n \in C(Q)$ (Q compact) is called a *Chebyshev system* (on Q) if every linear combination $\sum_{i=1}^n \alpha_i x_i \not\equiv 0$ has at most $n - 1$ zeros on Q. We have the following characterization theorem of E. Ya. Remez (see [168, p. 182]).

THEOREM 1.12. *Let $G = [x_1, \cdots, x_n]$ be an n-dimensional linear subspace of $E = C_R(Q)$ such that $x_1, \cdots, x_n$ form a Chebyshev system and let $x \in E \setminus G$ and $g_0 \in G$. We have $g_0 \in \mathscr{P}_G(x)$ if and only if there exist $n + 1$ distinct points $q_1, \cdots, q_{n+1} \in Q$ such that*

$$(1.40) \qquad x(q_j) - g_0(q_j) = \left(\operatorname{sign} \frac{\Delta_j}{\Delta}\right) \|x - g_0\|, \qquad j = 1, \cdots, n + 1,$$

where

$$(1.41) \qquad \Delta = \begin{vmatrix} x_1(q_1) \cdots x_1(q_{n+1}) \\ \cdots\cdots\cdots\cdots \\ x_n(q_1) \cdots x_n(q_{n+1}) \\ x(q_1) \cdots x(q_{n+1}) \end{vmatrix} \neq 0,$$

and where Δ_j is the cofactor of the element $x(q_j)$ in Δ.

This theorem follows as a particular case both from Theorem 1.6 for $\dim G = n < \infty$ and from Theorem 1.10 for $E = C_R(Q)$, using the general form (1.34) of the extremal points of S_{E^*} (naturally, since the scalars are real, we have now $\alpha = \pm 1$ in (1.34)). In particular, for $Q = [a, b]$ one obtains the following classical "alternation theorem" of P. L. Chebyshev–S. N. Bernstein (see [168, p. 184]).

THEOREM 1.13. *Let* $G = [x_1, \cdots, x_n]$ *be an n-dimensional linear subspace of* $E = C_R([a, b])$ *such that* $x_1, \cdots, x_n$ *form a Chebyshev system and let* $x \in E \setminus G$ *and* $g_0 \in G$. *We have* $g_0 \in \mathscr{P}_G(x)$ *if and only if there exist* $n + 1$ *points* $q_1 < q_2 < \cdots < q_{n+1}$ *of* $[a, b]$, *at which the difference* $x(q) - g_0(q)$ *takes the value* $\|x - g_0\|$ *with alternating signs (i.e., with opposite signs at consecutive points* $q_j, q_{j+1}, j = 1, \cdots, n$).

In $E = C(Q)$, taking into account (1.34), one obtains from Theorem 1.11 a result of E. W. Cheney [38, p. 73]; the necessity part of this latter result was observed, essentially, by T. J. Rivlin and H. S. Shapiro (see [168, p. 181]).

Some other characterizations of elements of best approximation by finite-dimensional subspaces in various concrete spaces are given in [168, pp. 201–206].

(f) A natural generalization of the problem of characterization of elements of best approximation is the problem of *simultaneous characterization* of a set of elements of best approximation: given E, $G \subset E$ and x as above and a subset M of G, what are the necessary and sufficient conditions in order that *every* element $g_0 \in M$ be an element of best approximation of x by the elements of G? The answer is given (see [168, p. 23]) by

THEOREM 1.14. *Let E be a normed linear space, G a linear subspace of E,* $x \in E \setminus \bar{G}$ *and* $M \subset G$. *We have* $M \subset \mathscr{P}_G(x)$ *if and only if there exists an* $f \in E^*$ *satisfying* (1.5), (1.6) *and*

$$f(x - g_0) = \|x - g_0\|, \qquad g_0 \in M. \tag{1.42}$$

In other words, this says that one can find the functional $f \in E^*$ of Theorem 1.1 to be common for all $g_0 \in M$. Theorem 1.14 is an immediate consequence of Theorem 1.1 and the observation that $\|x - g_1\| = \|x - g_2\|$ for all pairs $g_1, g_2 \in M \subset \mathscr{P}_G(x)$; naturally, the converse is also true, since Theorem 1.1 is even a particular case of Theorem 1.14. We shall see in § 3 that Theorem 1.14 has applications in the study of the uniqueness of elements of best approximation.

(g) Finally, let us mention two other tools for the study of problems of best approximation in normed linear spaces, which will be useful in the sequel. We mention them here since they also yield the "characterization theorems" (1.43) and (1.45) below.

Firstly, the canonical mapping $\omega_G : E \to E/G$ is clearly related to best approximation, since in a normed linear space E formula (1.3) can also be written in the form

$$\mathscr{P}_G(x) = \{g_0 \in G \mid \|x - g_0\| = \|\omega_G(x)\|\}. \tag{1.43}$$

Another useful tool in the study of best approximation will be the set

$$\mathscr{P}_G^{-1}(0) = \{x \in E \mid 0 \in \mathscr{P}_G(x)\}. \tag{1.44}$$

Some authors call the set $\mathscr{P}_G^{-1}(0)$ the *metric complement* of G (because of § 2, Proposition 2.1, § 3, Proposition 3.1 and § 4, Proposition 4.1).

For this set we have, obviously,

$$\mathscr{P}_G(x) = (x + \mathscr{P}_G^{-1}(0)) \cap G. \tag{1.45}$$

Some other properties of the set $\mathscr{P}_G^{-1}(0)$ are collected in

PROPOSITION 1.1. *Let G be a linear subspace of a normed linear space. E. Then*

(i) $0 \in \mathscr{P}_G^{-1}(0)$ *(hence* $\mathscr{P}_G^{-1}(0) \neq \emptyset$*).*

(ii) *The set* $\mathscr{P}_G^{-1}(0)$ *is closed.*

(iii) *If* $x \in \mathscr{P}_G^{-1}(0)$, *then* $\alpha x \in \mathscr{P}_G^{-1}(0)$ *for all scalars* α.

(iv) *If* $\dim E = \infty$, $\dim G < \infty$, *then* $\mathscr{P}_G^{-1}(0)$ *is homeomorphic to E.*

A relation between the mapping ω_G and the set $\mathscr{P}_G^{-1}(0)$ will be given in § 2, formula (2.10). In the sequel we shall see that various properties of best approximation can be characterized in terms of properties of ω_G (see, for example, § 2, Theorem 2.3), or of $\omega_G|_{\mathscr{P}_G^{-1}(0)}$ (see, for example, § 2, Proposition 2.1, § 3, Proposition 3.1, § 4, Theorems 4.2 and 4.18, etc.) or simply of the set $\mathscr{P}_G^{-1}(0)$ (see, for example, § 2, Proposition 2.1, § 3, Proposition 3.1, § 4, Proposition 4.1, Theorem 4.5, Proposition 4.7, etc.). Moreover, the set $\mathscr{P}_G^{-1}(0)$ is a useful tool not only for the study of best approximation, but also for other problems, e.g., for the problem of "simultaneous approximation and interpolation" (the so-called "property SAIN", see [50], [89]).

2. Existence of elements of best approximation.

2.1. Characterizations of proximinal linear subspaces.

(a) The basic notion in connection with the existence of elements of best approximation is the following.

DEFINITION 2.1. A set G in a metric space E is said to be *proximinal* if every element $x \in E$ has at least one element of best approximation in G, i.e., if

$$\mathscr{P}_G(x) \neq \emptyset, \qquad x \in E. \tag{2.1}$$

The term "proximinal" set (a combination of "proximity" and "minimal") was proposed by R. Killgrove and used first by R. R. Phelps [144, p. 790]. Some authors use for such sets the term *distance set*, or *existence set*, or *(E)-set*.

Since by § 1, formula (1.4), $\mathscr{P}_G(x) \neq \emptyset$ for all $x \in G$, condition (2.1) is equivalent to

$$\mathscr{P}_G(x) \neq \emptyset, \qquad x \in E \setminus G. \tag{2.2}$$

Let us also observe that every proximinal set is necessarily closed, since otherwise no $x \in \bar{G} \setminus G$ would have an element of best approximation in G.

(b) Now we shall consider the problem of characterization of proximinal linear subspaces G of a normed linear space E, i.e., the problem of giving necessary and sufficient conditions in order that G be proximinal. The basic observation is the following: *A linear subspace G of a normed linear space E is proximinal if and only if G is proximinal in every linear subspace* $E_0 \subset E$ *such that G is a closed hyperplane in* E_0 (i.e., a closed linear subspace of E_0 such that $\dim E_0/G = 1$). In fact, although this observation is obvious, it is useful since it reduces the problem of characterization of proximinal linear subspaces to that of the characterization of proximinal closed hyperplanes. This latter problem is solved by the observation that *a closed hyperplane G in* E_0 *is proximinal if and only if there exists an element* $z \in E_0 \setminus \{0\}$

such that $0 \in \mathscr{P}_G(z)$. (Indeed, if $x \in E_0 \setminus G$ and $g_0 \in \mathscr{P}_G(x)$, then $0 \in \mathscr{P}_G(x - g_0)$ and conversely, if $0 \in \mathscr{P}_G(z)$ and $x \in E_0 \setminus G$, then, since G is a closed hyperplane in E_0, we have $x - \alpha z \in G$ for suitable scalar $\alpha \neq 0$ whence, since also $0 \in \mathscr{P}_G(\alpha z)$, we infer $x - \alpha z \in \mathscr{P}_G(\alpha z + x - \alpha z) = \mathscr{P}_G(x)$; note that this proof is somewhat shorter than that given in [168, pp. 93–94]. Using these observations, and the characterization of best approximations given in § 1, Theorem 1.1, we obtain the following main theorem of characterization of proximinal linear subspaces.

THEOREM 2.1. *A linear subspace G of a normed linear space E is proximinal if and only if G is closed and for every linear subspace $E_0 \subset E$ such that G is a closed hyperplane in E_0 and every functional $\varphi \in E_0^*$ with $\varphi|_G = 0$, there exists an element $z \in E_0 \setminus \{0\}$ such that*

$$\varphi(z) = \|\varphi\|, \quad \|z\| = 1. \tag{2.3}$$

Any element z satisfying (2.3) is called a *maximal element* of φ. In the usual concrete normed linear spaces the general form of functionals which admit maximal elements and the general form of maximal elements of such a functional are well known and simple (see for example, [154], [198]) and therefore Theorem 2.1 is suitable for applications to certain subspaces in concrete spaces.

From § 1, formula (1.23) and $(G^\perp)^* \equiv (E/G)^{**}$ one obtains the following characterizations of proximinal subspaces, the second of which is due to A. L. Garkavi (see [168, pp. 94–95]).

THEOREM 2.2. *For a closed linear subspace G of a normed linear space E the following statements are equivalent:*

1°. *G is proximinal.*

2°. *For every $x \in E$ there exists an element $y \in E$ such that*

$$f(y) = f(x), \qquad f \in G^\perp, \tag{2.4}$$

$$\|y\| = \|\kappa(x)|_{G^\perp}\|. \tag{2.5}$$

If E/G is reflexive, these statements are equivalent to:

3°. *For every $\Phi \in (G^\perp)^*$ there exists an element $y \in E$ such that*

$$f(y) = \Phi(f), \qquad f \in G^\perp, \tag{2.6}$$

$$\|y\| = \|\Phi\|. \tag{2.7}$$

Let us also mention the following two characterizations of proximinal linear subspaces, the first of which appears in Cheney-Wulbert [46].

PROPOSITION 2.1. *For a linear subspace G of a normed linear space E the following statements are equivalent:*

1°. *G is proximinal.*

2°. *We have*

$$E = G + \mathscr{P}_G^{-1}(0) = \{g + x \mid g \in G, x \in \mathscr{P}_G^{-1}(0)\}. \tag{2.8}$$

3°. *G is closed and for the canonical mapping $\omega_G : E \to E/G$ we have*

$$\omega_G(\mathscr{P}_G^{-1}(0)) = E/G. \tag{2.9}$$

(In other words, $\omega_G|_{\mathscr{P}_G^{-1}(0)}$ maps $\mathscr{P}_G^{-1}(0)$ onto E/G.)

Indeed, if G is proximinal and $x \in E$, $g_0 \in \mathscr{P}_G(x)$, then $x = g_0 + (x - g_0) \in G + \mathscr{P}_G^{-1}(0)$. Conversely, if we have (2.8) and $x \in E$, $x = g_0 + y$, where $g_0 \in G$, $y \in \mathscr{P}_G^{-1}(0)$, then $0 \in \mathscr{P}_G(y) = \mathscr{P}_G(x - g_0)$, whence $g_0 \in \mathscr{P}_G(x)$. Thus, $1° \Leftrightarrow 2°$. Furthermore, if G is proximinal and $x + G \in E/G$, $g_0 \in \mathscr{P}_G(x)$, then $x - g_0 \in \mathscr{P}_G^{-1}(0)$ and $\omega_G(x - g_0) = x + G$. Conversely, if we have 3° and $x \in E$, then $x + G \in E/G = \omega_G(\mathscr{P}_G^{-1}(0))$, so $x + G = \omega_G(y)$, where $y \in \mathscr{P}_G^{-1}(0)$. Hence $x - y = g_0 \in G$ and $\|x - g_0\| = \|y\| = \inf_{g \in G} \|y - g\| = \inf_{g \in G} \|x - g_0 - g\|$, so $g_0 \in \mathscr{P}_G(x)$. Thus, $1° \Leftrightarrow 3°$, which completes the proof.

In connection with 3° above let us observe that for any closed linear subspace $G \in E$ we have, by definition, $\|\omega_G(x)\| = \rho(x, G)$ $(x \in E)$ and hence, by § 1, formula (1.44),

$$\mathscr{P}_G^{-1}(0) = \{x \in E \,|\, \|\omega_G(x)\| = \|x\|\}; \tag{2.10}$$

in other words, $\mathscr{P}_G^{-1}(0)$ is that subset of E, on which the restriction of ω_G is *norm-preserving*.

We have also the following useful characterizations of proximinality in terms of the canonical mapping $\omega_G : E \to E/G$ and of the unit cells $S_E = \{x \in E \,|\, \|x\| \leqq 1\}$, $S_{E/G} = \{x + G \in E/G \,|\, \|x + G\| \leqq 1\}$, due to G. Godini [74].

THEOREM 2.3. *For a closed linear subspace G of a normed linear space E the following statements are equivalent*:

1°. *G is proximinal.*

2°. *$\omega_G(S_E)$ is closed in E/G.*

3°. *We have*

$$\omega_G(S_E) = S_{E/G}. \tag{2.11}$$

(c) Some characterizations of proximinal subspaces of finite codimension in general normed linear spaces are given in [176] and [74]. We mention here the following ones from [176].

THEOREM 2.4. *Let E be normed linear space and G a subspace of codimension n of E (i.e., $\dim E/G = n$). The following statements are equivalent*:

1°. *G is proximinal.*

2°. *There exists a basis $\{f_1, \cdots, f_n\}$ of $G^\perp$ such that the set*

$$A = \{\{f_i(y)\}_1^n \,|\, y \in S_E\} \tag{2.12}$$

is closed in the n-dimensional Euclidean space H_n.

3°. *For every basis $\{f_1, \cdots, f_n\}$ of $G^\perp$ the set A is closed in H_n.*

If, in addition, every bounded closed convex set $\mathscr{C}$ in E is the closed convex hull of the set $\mathscr{E}(\mathscr{C})$ of its extremal points (for example, [7], *if E is isomorphic to a subspace of a separable conjugate space E_0^*), then in* 2° *and* 3° *one can replace A by the set*

$$D = \operatorname{co}\{\{f_i(x)\}_1^n \,|\, x \in \mathscr{E}(S_E)\} = \{\{f_i(y)\}_1^n \,|\, y \in \operatorname{co}\mathscr{E}(S_E)\}, \tag{2.13}$$

where co *stands for "convex hull".*

(d) Let us give now some applications of the above theorems to characterizations of proximinal linear subspaces of finite codimension in concrete spaces. Using

Theorem 2.2, equivalence $1^\circ \Leftrightarrow 3^\circ$, A. L. Garkavi has proved (see [168, p. 302]):

THEOREM 2.5. *A closed linear subspace G of codimension n of $E = C_R(Q)$ (Q compact) is proximinal if and only if the following three conditions are satisfied:*

(i) *For every $\mu \in G^\perp \setminus \{0\}$ the carrier $S(\mu)$ admits a Hahn-decomposition into two closed sets $S(\mu)^+$ and $S(\mu)^- = S(\mu) \setminus S(\mu)^+$.*

(ii) *For every pair of measures $\mu_1, \mu_2 \in G^\perp \setminus \{0\}$ the set $S(\mu_1) \setminus S(\mu_2)$ is closed.*

(iii) *For every pair of measures $\mu_1, \mu_2 \in G^\perp \setminus \{0\}$ the measure μ_1 is absolutely continuous with respect to μ_2 on the set $S(\mu_2)$.*

In the particular case when $n = 1$ (i.e., when G is a closed hyperplane), conditions (ii) and (iii) are automatically satisfied (since $\dim G^\perp = 1$) and hence condition (i) is necessary and sufficient in order that G be proximinal; one can also show directly that this condition is equivalent to that of Theorem 2.1, i.e., to the existence of a maximal element for each $\mu \in G^\perp \setminus \{0\}$.

If $E = c_0$, from Theorem 2.4 one obtains [176] the following characterizations of proximinal subspaces of finite codimension, due to J. Blatter and E. W. Cheney [14] (see also W. Pollul [149, Corollary 2.7]) and W. Pollul [149, Lemma 2.6] respectively.

THEOREM 2.6. *For a closed linear subspace G of codimension n of $E = c_0$ the following statements are equivalent:*

1°. *G is proximinal.*

2°. *For every $f \in G^\perp \setminus \{0\}$ we have*

$$\mathcal{M}_f = \{x \in E \mid f(x) = \|f\|, \|x\| = 1\} \neq \emptyset. \tag{2.14}$$

3°. *There exists a basis $\{f_1, \cdots, f_n\}$ of $G^\perp$ such that each f_i, $i = 1, \cdots, n$, satisfies* (2.14).

If $E = l^1$, from Theorem 2.4 one obtains the following result [176] (in particular, for real scalars, see A. L. Garkavi [70, Theorem 1] and [71, Theorem 5]).

THEOREM 2.7. *For a closed linear subspace G of codimension n of $E = l^1$ the following statements are equivalent:*

1°. *G is proximinal.*

2°. *There exists a basis $\{f_1, \cdots, f_n\}$ of $G^\perp$ such that (or, equivalently, for every basis $f_1, \cdots, f_n$ of $G^\perp$) the set*

$$D = \operatorname{co} \{\varepsilon\{f_i(e_j)\}_1^n \mid |\varepsilon| = 1, j = 1, 2, \cdots\} \tag{2.15}$$

is closed in the n-dimensional Euclidean space H_n (where $e_j = \{\underbrace{0, \cdots, 0}_{j-1}, 1, 0, \cdots\}$).

Some characterizations of proximinal subspaces of finite codimension in $E = L_R^1(T, \nu)$, where (T, ν) is a positive measure space such that $L_R^1(T, \nu)^* \equiv L_R^\infty(T, \nu)$, have been given by A. L. Garkavi [71].

2.2. Some classes of proximinal linear subspaces.

(a) Whenever a new class (=family) of subspaces is introduced, it is natural to ask whether it is nonvoid; in particular, one can ask whether proximinal linear subspaces exist in every normed linear space. The answer is affirmative, since from Theorem 2.1 it follows that whenever $f \in E^*$ has a maximal element, $G = \{x \in E | f(x) = 0\}$ is a proximinal hyperplane.

Problem 2.1. Let $1 < n < \infty$. Does every normed linear space (or, in particular, every Banach space) E contain a proximinal subspace G of codimension n?

This problem has been raised in [176].

From Theorem 2.9 (i) below it follows that if E is linearly isometric to a conjugate Banach space, the answer is affirmative. By Theorem 2.4, Problem 2.1 amounts to finding a subspace G of E such that for any basis $\{f_1, \cdots, f_n\}$ of $G^{\perp}$ the set (2.12) is closed in H_n. Using Theorem 2.5, A. L. Garkavi has proved (see [168, p. 310]) that *for any compact space Q and any integer n with $1 \leqq n < \dim C_R(Q)$, the space $E = C_R(Q)$ contains proximinal subspaces G of codimension n*, for example, if $q_1, \cdots, q_n \in Q$, then $G = \{x \in C_R(Q) | x(q_i) = 0 \ (i = 1, \cdots, n)\}$ is such a subspace. Also, A. L. Garkavi has proved [71] that *for any positive measure space (T, ν) such that $L_R^1(T, \nu)^* \equiv L_R^{\infty}(T, \nu)$ and any integer n with $1 \leqq n < \dim L_R^1(T, \nu)$, the space $E = L_R^1(T, \nu)$ contains proximinal subspaces G of codimension n*, for example, if $A_1, \cdots, A_n$ are disjoint sets with $\nu(A_i) > 0$, $i = 1, \cdots, n$, then $G = \{x \in L_R^1(T, \nu) | \int_{A_i} x(t)\, d\nu(t) = 0 \ (i = 1, \cdots, n)\}$ is such a subspace.

(b) Now we shall give some other important classes of proximinal linear subspaces, involving compactness in weak and weak* topologies.

THEOREM 2.8. *Let E be a normed linear space and let G be a linear subspace of E such that the unit cell $S_G = \{g \in G | \|g\| \leqq 1\}$ is sequentially compact for the weak topology $\sigma(E, E^*)$. Then G is proximinal.*

This theorem (due to V. Klee [102]) can be deduced from Theorem 2.1 but it admits also a simple direct proof; a similar remark is also valid for Theorem 2.9 below. For the proofs, see [168, Chap. I, § 2]. An immediate consequence of Theorem 2.8 is the following corollary.

COROLLARY 2.1. *Let E be a normed linear space and let G be a linear subspace of E with the property that G is a reflexive Banach space. Then G is proximinal. In particular, every finite-dimensional linear subspace G of a normed linear space E is proximinal.*

THEOREM 2.9. *Let E^* be the conjugate space of a normed linear space E. Then*

(i) *Every linear subspace Γ of E^* having the unit cell $S_{\Gamma} = \{f \in \Gamma | \|f\| \leqq 1\}$ compact for $\sigma(E^*, E)$ is proximinal. In particular, every $\sigma(E^*, E)$-closed linear subspace Γ of E^* is proximinal.*

(ii) *Every linear subspace Γ of E^* having the unit cell S_{Γ} sequentially compact for $\sigma(E^*, E)$ is proximinal.*

Note that the first statement in (i) is indeed more general than the second, since we did not assume E complete. Also, it can be shown by examples that between (i) and (ii) there is no relation of implication.

The conditions of Theorems 2.8 and 2.9 are sufficient, but not necessary, in order that G or Γ be proximinal in E or E^* respectively. Indeed, if E is nonreflexive and $x \in E \setminus \{0\}$, then for any $f \in E^* \setminus \{0\}$ such that $f(x) = \|f\| \, \|x\|$, the hyperplane $G = \{x \in E | f(x) = 0\}$ is proximinal in E, but S_G is not weakly sequentially compact. Also, if E is nonreflexive and $f \in E^* \setminus \{0\}$ has no maximal element in E, then for any $\Phi \in E^{**} \setminus \{0\}$ such that $\Phi(f) = \|\Phi\| \, \|f\|$, the hyperplane $\Gamma = \{\gamma \in E^* | \Phi(\gamma)$

$= 0\}$ is proximinal in E^*, but Γ is $\sigma(E^*, E)$-dense in E^* and hence S_Γ is not $\sigma(E^*, E)$-compact; if E is separable, it follows that S_Γ is not sequentially compact for $\sigma(E^*, E)$ (for example, by [55, p. 426, Theorem 1]). Some other important examples of $\sigma(E^*, E)$-dense proximinal subspaces of conjugate spaces E^* will be given below (see the results on certain proximinal subspaces of $l_R^\infty(\Omega)$ and the results of § 2.5).

Let us also mention the following partial extension of the second statement of Theorem 2.9 (i) to spaces of continuous linear operators, given by P. Franck [65].

THEOREM 2.10. *Let E_1, E_2, be two normed linear spaces and G_1 a linear subspace of E_1 such that there exists a projection p of norm $\|p\| = 1$ from E_1 onto G_1. Then the subspace*

$$\Gamma = \{v \in L(E_1, E_2) | v(y) = 0 \ (y \in G_1)\} \tag{2.16}$$

of $L(E_1, E_2)$ (of all continuous linear mappings of E_1 into E_2, with the usual norm) is proximinal.

In particular, if E_1 is an inner product space, then every linear subspace G_1 of E_1 satisfies the hypothesis of Theorem 2.10.

(c) In some concrete spaces E, for certain linear subspaces G of infinite dimension and infinite codimension, which do not satisfy one of the above sufficient conditions for proximinality (of Theorem 2.8 or Theorem 2.9), and for which E/G is non-reflexive, the general characterizations given in Theorems 2.1–2.3 and Proposition 2.1 are not suitable to decide whether G is proximinal or not and each such case must be handled with different ad hoc methods. Let us now give some examples.

For a compact space Q and a Banach space F we shall denote by $C_F(Q)$ the space of all functions defined on Q and with values in F, strongly continuous on Q (i.e., $\lim_{q \to q_0} \|z(q) - z(q_0)\|_F = 0$ for all $z(\cdot) \in C_F(Q)$), endowed with the usual vector operations and with the norm $\|z(\cdot)\| = \max_{q \in Q} \|z(q)\|_F$; here we use the notation $z(\cdot)$ for the elements of $C_F(Q)$, in order to distinguish them from the elements $z \in F$. A. Pelczynski [142] has raised the following problem. Let Q, S be two compact spaces, $\psi^0 : C_F(S) \to C_F(Q)$ the "conjugate" mapping defined by

$$\psi^0(y(\cdot)) = y(\cdot) \circ \psi, \qquad y(\cdot) \in C_F(S); \tag{2.17}$$

is the closed linear subspace $G = \psi^0(C_F(S))$ of $E = C_F(Q)$ proximinal in $C_F(Q)$? An affirmative answer has been given by C. Olech [134] for uniformly convex F and by J. Blatter [13] for real Banach spaces F such that the conjugate space F^* is linearly isometric to $L^1(T, v)$ for some positive measure space (T, v); they have also given a formula for the distance $\rho(y(\cdot), G)$, $y(\cdot) \in E$. In the particular case when $F = R$, the proximinality of G was established by S. Mazur (unpublished) and the formula for $\rho(y(\cdot), G)$ was announced, without proof, in [142] and proved also by S. Mazur (unpublished); Mazur's proofs can be found in [158, p. 20] and [159, p. 124]. In the particular case when $F = C$ (the field of complex scalars), the formula for $\rho(y(\cdot), G)$ has been conjectured by Z. Semadeni and proved by A. Pelczynski [143].

More generally, J. Blatter [13] has studied the problem of proximinality of Grothendieck subspaces G of $C_F(Q)$ spaces, i.e., of subspaces G consisting of all functions in $C_F(Q)$ satisfying a set I of relations of the form

(2.18) $$x(q_i) = \alpha_i x(q_i'), \quad \text{where} \quad q_i, q_i' \in Q, \quad \alpha_i = \text{scalar}, \quad i \in I;$$

for example, he has proved [13]

THEOREM 2.11. *If G is a Grothendieck subspace of $E = C_F(Q)$, where Q is compact and $F^* \equiv L^1(T, \nu)$ for some positive measure space (T, ν), and if*

(2.19) $$\lambda = \inf_{\substack{q \in Q \setminus \cap Z(g) \\ g \in G}} \sup_{\substack{g \in G \\ \|g\| = 1}} \|g(q)\| > 0,$$

where $Z(g) = \{q \in Q | g(q) = 0\}$, then G is proximinal in $E = C_F(Q)$.

In particular, if G is a Weierstrass-Stone subspace of $C_F(Q)$, i.e., [13], if G consists of all functions in $C_F(Q)$ satisfying (2.18) with $\alpha_i = 1$ or 0 for all $i \in I$, then

$$\sup_{\substack{g \in G \\ \|g\| = 1}} \|g(q)\| = 1$$

for each $q \in Q \setminus \bigcap_{g \in G} Z(g)$, hence G satisfies (2.19) and thus one obtains the following corollary [13].

COROLLARY 2.2. *If G is a Weierstrass-Stone subspace of $E = C_F(Q)$, where Q is compact and $F^* \equiv L^1(T, \nu)$, then G is proximinal in $E = C_F(Q)$.*

Note that the subspaces $G = \psi^0(C_F(S))$ of $C_F(Q)$, considered in the abovementioned problem of A. Pelczynski, are Weierstrass-Stone subspaces of $C_F(Q)$ (consisting of all functions which are constant on each $\psi^{-1}(s)$), which contain all constant mappings of Q into F. In the particular case when $F =$ the field of scalars, a subspace G of $C(Q)$ is a Weierstrass-Stone subspace of $C(Q)$ if and only if G is a self-adjoint closed subalgebra of $C(Q)$ (see, for example, [13]); such a subspace G of $C(Q)$ is of the form $G = \psi^0(C(S))$ for a suitable compact space S and a continuous mapping ψ of Q onto S if and only if G contains the constant functions (see, for example, [159, p. 122, Theorem 7.5.2]).

Independently, R. B. Holmes and B. R. Kripke have proved (see [85], [86] and [82]), using interposition of continuous functions between two other functions, that if Ω is a paracompact (e.g., a metric) space and $E = l_R^\infty(\Omega)$, then the subspace $G = l_R^\infty(\Omega) \cap C_R(\Omega)$ is proximinal in E (namely, they have used the observation that a subset G of $E = l_R^\infty(\Omega)$ is proximinal if and only if it has the following "interposition property": for each $x \in l_R^\infty(\Omega)$ there exists a $g_0 \in G$ such that

$$\lim_{n \to \infty} \inf_{\substack{g \in G \\ g \geqq x - \rho(x,G) - 1/n}} g(t) \leqq g_0(t) \leqq \lim_{n \to \infty} \sup_{\substack{g \in G \\ g \leqq x + \rho(x,G) + 1/n}} g(t)$$

for all $t \in \Omega$); they have also computed $\rho(x, G)$, $x \in E$. However, it turns out that this is a particular case of the abovementioned result of S. Mazur. Indeed, it is well known that $E = l_R^\infty(\Omega)$ is linearly isometric to a $C_R(Q)$ space, where Q is compact (see, for example, [55]) and clearly $G = l_R^\infty(\Omega) \cap C_R(\Omega)$ is a closed subalgebra of E, containing the constants, whence by the above remarks G is a Weierstrass-Stone subspace of $C_R(Q)$, of the form $\psi^0(C_R(S))$. Let us also mention that C. Olech [134] and J. Blatter [13] have proved, more generally, that if S is a compact space and F a uniformly convex Banach space (respectively, a Banach space such that $F^* \equiv L^1(T, \nu)$), then every upper semi-continuous mapping $\mathscr{U}$ (see § 4 for the definition of such mappings) of S into the family of all closed subsets of F (respectively, of all nonvoid compact subsets of F) has an element of best approximation in $G = C_F(S)$ (respectively, in Grothendieck subspaces G of $C_F(S)$ satisfying certain conditions), in the sense of the "distance" defined by

$$\rho(\mathscr{U}, y(\cdot)) = \sup_{s \in S} \sup_{y' \in \mathscr{U}(s)} \|y(s) - y'\|_F, \qquad y(\cdot) \in C_F(S); \tag{2.20}$$

actually, C. Olech [134] has applied this result to derive his solution of Pelczynski's problem mentioned above, by observing that if $z(\cdot) \in C_F(Q)$, then

$$\mathscr{U}(s) = \{z(q) | q \in \psi^{-1}(s)\} \tag{2.21}$$

defines an upper semi-continuous mapping of S into the family of all closed subsets of F and that if $y_0(\cdot) \in C_F(S)$ satisfies $y_0(\cdot) \in \mathscr{P}_{C_F(S)}(\mathscr{U})$, then $g_0(\cdot) = \psi^0(y_0(\cdot)) \in \psi^0(C_F(S)) = G$ satisfies $g_0(\cdot) \in \mathscr{P}_G(z(\cdot))$ in $E = C_F(Q)$. However, the set of all upper semi-continuous mappings $\mathscr{U}$ of S into the family of closed subsets of F is not linear. Note that $\sup_{y' \in \mathscr{U}(s)} \|y(s) - y'\|_F$ in (2.20) is nothing else than the distance between the subsets $\{y(s)\}$ (=singleton) and $\mathscr{U}(s)$ of F with respect to the Hausdorff metric in 2^F (see § 4). In this connection we mention also the work of B. Sendov (e.g., [160], [161]), who has defined the distance between two functions on $Q = [a, b]$ as the Hausdorff distance (with respect to a given Minkowski distance in the plane R^2) between their "completed graphs" (the completed graph of a real function $x(\cdot)$ on $[a, b]$ is the intersection of all closed sets A in R^2 containing the graph of $x(\cdot)$ and such that the relations $\{t, y\}, \{t, z\} \in A$, $t < u < z$, imply $\{t, u\} \in A$) and has studied problems of best approximation for this distance; however, in general the usual concrete linear spaces of functions endowed with this metric are not Banach spaces.

Another problem which has been studied in the spaces $C_F(Q)$ is that of "best approximation by operator functions" (see the references given in [168, p. 226]): Let Q be a compact space, F_1, F two Banach spaces, and $u(\cdot)$ a function defined on Q whose values are closed linear operators from F_1 into F, having the same domain of definition $\mathscr{D}_{u(q)} = \mathscr{D}$ $(q \in Q)$, dense in F_1, and satisfying the following two hypotheses: (i) for every $x \in \mathscr{D}$ the function $g_x(\cdot)$: $q \to u(q)x$ belongs to $C_F(Q)$; (ii) the relations $x \in \mathscr{D}$, $g_x(q) = u(q)x = 0$ $(q \in Q)$ imply $x = 0$. S. I. Zukhovitskii and G. I. Eskin [199] have proved that *if F_1 and F are reflexive, the linear subspace $G = \{g_x(\cdot) | x \in \mathscr{D}\}$ of $C_F(Q)$ is proximinal if and only if there exists a constant $m > 0$ such that* $\max_{q \in Q} \|u(q)x\|_F \geqq m\|x\|_{F_1}$ *for all $x \in \mathscr{D}$*; in the particular case when $u(q) \in L(F_1, F)$ $(q \in Q)$, this result has been obtained by S. B. Stechkin [178].

Several papers have also been devoted to the study of best approximation of bounded functions by bounded linear functions. Thus, B. R. Kripke and R. T. Rockafellar [111] have proved the necessity part and E. W. Cheney and A. A. Goldstein [40] the sufficiency part of the following result.

THEOREM 2.12. *Let F be a real Banach space and Ω a bounded subset of F. The subspace $G = F^*|_\Omega$ is proximinal in $E = l_R^\infty(\Omega)$ if and only if 0 is interior to the closed convex hull of $\Omega \cup (-\Omega)$ relative to its linear span.*

R. B. Holmes and B. R. Kripke [86] have extended the sufficiency part to real locally convex spaces, in terms of their "interposition property" mentioned above.

A somewhat stronger condition on Ω is already sufficient for proximinality in some more general spaces. Namely, E. W. Cheney and A. A. Goldstein [40] have proved that *if F_1, F_2 are two real Banach spaces, $\Omega \subset F_1$ and 0 is in the interior of the closed convex hull of $\Omega \cup (-\Omega)$, then the subspace $G = L(F_1, F_2^*)|_\Omega$ of $E = l^\infty_{F_2^*}(\Omega)$ is proximinal.* In the particular case when $\Omega = S_{F_1}$ (which clearly satisfies the condition), the proximinality of G in E has been proved by A. L. Garkavi [66]. The above result has been further extended by Y. Ikebe [92], who has shown that *if F_1, F_2 are two (real or complex) normed linear spaces and 0 is in the interior of the closed convex circled hull of $\Omega \subset F_1$, then the subspace $G = L(F_1, F_2^*)|_\Omega$ is proximinal in the space E of all bounded mappings of Ω into $2^{F_2^*}$, endowed with the natural vector operations (namely, $(\mathscr{U} - \mathscr{V})(x) = \mathscr{U}(x) - \mathscr{V}(x) = \{y - z \mid y \in \mathscr{U}(x),\ z \in \mathscr{V}(x)\}$) and with the norm $\|\mathscr{U}\| = \sup_{h \in \cup_{y \in \Omega} \mathscr{U}(y)} \|h\|$* (a mapping $\mathscr{U} : \Omega \to 2^{F_2^*}$ is said to be bounded if $\|\mathscr{U}\| < \infty$; the space E of all these $\mathscr{U}$ is a Banach space). Let us also mention that Y. Ikebe [92] has shown the necessity of the condition of Theorem 2.12 in the following more general setting: *If the subspace $G = L(F_1, F_2)|_\Omega$ of $l^\infty_{F_2}(\Omega)$ is proximinal, then 0 is interior to the closed convex hull of Ω relative to its linear span.*

Some more classes of proximinal subspaces of infinite dimension and infinite codimension have been obtained in spaces of continuous linear operators (cf. Theorem 2.10 above). Thus, I. C. Gohberg and M. G. Krein [75, Chap. II, § 7, Corollary 7.1] have proved

THEOREM 2.13. *Let H be a Hilbert space (i.e., a complete inner product space). Then the subspace $G = \mathscr{C}(H, H)$ of all compact linear mappings of H into H is proximinal in the space $E = L(H, H)$.*

The distance $\rho(u, G)$, $u \in E$, has also been computed in [75]. Theorem 2.13 has been reproved, independently, by R. B. Holmes and B. R. Kripke [88], who have extended it to $E = L(H_1, H_2)$, where H_1, H_2 are two different Hilbert spaces. Let us observe here that one can also give a third, rather simple, proof of Theorem 2.13, as follows: It is known (see, for example, [47, p. 52, Lemma 4.1 with $a = 1$, $b = 1$]) that *if G is a two-sided ideal in a C^*-algebra E, then* G satisfies condition 3° of Theorem 2.3, whence *G is proximinal*; it is also well known that $\mathscr{C}(H, H)$ is a two-sided ideal in the C^*-algebra $L(H, H)$, which completes the proof.

R. B. Holmes and B. R. Kripke [88] have observed that there are also other Banach spaces F such that $G = \mathscr{C}(F, F)$ is proximinal in $E = L(F, F)$ (actually, their example F is isomorphic to a Hilbert space) and that there exist Banach spaces F for which no $u \in L(F, F)$ has an element of best approximation in $G = \mathscr{C}(H, H)$ (again, their example F is isomorphic to a Hilbert space). Therefore, they have raised the following problem [88].

Problem 2.2. Characterize those Banach spaces F for which $G = \mathscr{C}(F, F)$ is proximinal in $E = L(F, F)$.

We shall make a remark on this problem in § 2.5 below.

Finally, let us mention the following result of Ky Fan and A. Hoffman [63]: *If H is a Hilbert space, then the subspace $G = \mathscr{H}(H, H)$ of all Hermitian operators* *is proximinal in* $E = L(H, H)$; namely, for each $u \in E = L(H, H)$ we have $(u + u^*)/2 \in \mathscr{P}_G(u)$.

Some more classes of proximinal linear subspaces will be given in § 2.5 below.

2.3. Normed linear spaces in which all closed linear subspaces are proximinal. Along with every new class of linear subspaces of normed linear spaces which we introduce, it is natural to consider also the complementary class, i.e., the family of all linear subspaces which do not belong to that class; we shall denominate the linear subspaces of this family by the prefix "non-" followed by the name of the original class, e.g., the linear subspaces which are not proximinal will be called "non-proximinal". Naturally, in every infinite-dimensional normed linear space there exist non-proximinal linear subspaces, for example, the nonclosed linear subspaces. However, now we shall show that the problem of existence of non-proximinal *closed* linear subspaces may have a negative answer, i.e., there exist normed linear spaces in which all *closed* linear subspaces are proximinal and we shall give some characterizations of such spaces.

THEOREM 2.14. *For a normed linear space E the following statements are equivalent*:

1°. *All closed linear subspaces of E are proximinal.*

2°. *The restriction of each* $f \in E^*$ *to every closed linear subspace of E has a maximal element.*

If E is a Banach space, these statements are equivalent to the following:

3°. *All closed linear subspaces of E, of a certain fixed finite codimension m, where* $1 \leqq m < \dim E$, *are proximinal.*

4°. *Every* $f \in E^*$ *has a maximal element.*

5°. *E is reflexive.*

The equivalence 1° $\Leftrightarrow$ 2° follows from Theorem 2.1, and the other equivalences are a consequence of the profound theorem of R. C. James [97], [99] (for which only difficult proofs are known to-day) that 4° $\Rightarrow$ 5°. Note that a normed linear space E satisfying 4° need not be a Banach space (and thus it need not satisfy 5°) as shown by the following example of R. C. James [98]: Let B be the space of all sequences of real numbers

$$x = \{\xi_1^1; \xi_1^2, \xi_2^2; \xi_1^3, \xi_2^3, \xi_3^3; \cdots\} \tag{2.22}$$

such that

$$\|x\| = \{|\xi_1^1|^2 + [\max(|\xi_1^2|, |\xi_2^2|)]^2 + [\max(|\xi_1^3|, \xi_2^3|, |\xi_3^3|)]^2 + \cdots\}^{1/2} < \infty, \tag{2.23}$$

endowed with the usual operations and with the norm (2.23), and let E be the linear span of all members x of B for which

$$|\xi_1^n| = |\xi_2^n| = \cdots = |\xi_n^n| \quad \text{for } n = 1, 2, \cdots. \tag{2.24}$$

Then B is a reflexive Banach space and E is a dense linear subspace of B with $E \neq B$ (hence E is not a Banach space), having the property 4° of Theorem 2.14 above.

2.4. Normed linear spaces which are proximinal in every superspace. We have seen in Corollary 2.1 above that if G is a reflexive Banach space, then G is proximinal in every superspace E (i.e., in every normed linear space E containing G as a subspace). It is natural to raise the problem of characterization of all normed linear spaces G with this property. It is obvious that a normed linear space G which is proximinal in every superspace must be complete, i.e., a Banach space, since every noncomplete normed linear space G is nonproximinal in its completion. Recently W. Pollul [149] has proved that each nonreflexive Banach space G can be embedded isometrically as a nonproximinal closed hyperplane in another Banach space E. Thus we have

THEOREM 2.15. *A normed linear space G is proximinal in every superspace E if and only if G is a reflexive Banach space.*

The proof of Pollul [149] has used James's theorem (the implication $4° \Rightarrow 5°$ of Theorem 2.14 above), by observing that if $\varphi \in G^*$ has no maximal element, then $\{G, 0\}$ is a nonproximinal hyperplane in $E = G \times R$ endowed with the norm $\|\{x, \lambda\}\| = \max(\|x\|, |\varphi(x) + \lambda|)$ and G is isometric to $\{G, 0\}$. A more elementary proof of Theorem 2.15, which does not make use of James' theorem, was given recently in [172].

2.5. Normed linear spaces which are, under the canonical embedding, proximinal in their second conjugate space.

(a) J. Blatter [10], [13] has raised

Problem 2.3. For which normed linear spaces E is $\kappa(E)$ proximinal in E^{**}?

Following R. B. Holmes and J. D. Ward [90], we shall say that such a normed linear space has *property proxbid* (a combination of the words "proximinal" and "bidual"). Clearly, every normed linear space with property proxbid is complete, i.e., a Banach space. The first result concerning Problem 2.3 is actually Theorem 2.13 above, which shows that *if H is a Hilbert space, then $G = \mathscr{C}(H, H)$ has property proxbid*; indeed, it is well known (see, for example, [157]) that $G^{**} = \mathscr{C}(H, H)^{**}$ is linearly isometric to $L(H, H)$ and that the image of $\kappa(\mathscr{C}(H, H))$ under this isometry is $\mathscr{C}(H, H)$. Also, the remarks made after Theorem 2.13 show that there exist other Banach spaces E which have property proxbid and that *$G = \mathscr{C}(H, H)$ admits an equivalent norm $|||\cdot|||$ such that if $E = (G, |||\cdot|||)$, then no $\Phi \in E^{**} \setminus \kappa(E)$ has an element of best approximation in $\kappa(E)$.*

Some other important classes of spaces with property proxbid are given in the following theorem.

THEOREM 2.16. (i) *If Q is a compact space, then $E = C(Q)$ has property proxbid.*

(ii) *If T is a locally compact space, then $E = C_0(T)$ has property proxbid.*

(iii) *If (T, ν) is a positive measure space, then $E = L_R^1(T, \nu)$ has property proxbid.*

Here $C_0(T)$ denotes the space of all (real or complex) continuous functions x on T which vanish at infinity (i.e., for each $\varepsilon > 0$ there exists a compact subset Q of T such that $|x(t)| < \varepsilon$ for all $t \in T \setminus Q$), endowed with the usual vector operations and with the norm $\|x\| = \sup_{t \in T} |x(t)|$. Part (i) of Theorem 2.16 has been proved by J. Blatter and G. Seever [16] and, independently, with a different method, by R. B. Holmes and J. D. Ward [90]. For real scalars, part (ii) has been proved by

J. Blatter [13] and for complex scalars, differently, by R. B. Holmes and J. D. Ward [90]; for both parts, the elements of best approximation (i.e., of $\mathscr{P}_{\kappa(E)}(\Phi)$, where $\Phi \in E^{**} \setminus \kappa(E)$), have been characterized in [90] as the continuous selections of a certain lower semi-continuous mapping (see § 4) and the distance $\rho(\Phi, \kappa(E))$, $\Phi \in E^{**}$, has been evaluated.

In view of Theorems 2.13 and 2.16 (i), (ii), R. B. Holmes and J. D. Ward [90] have raised

Problem 2.4. Does every C^*-algebra have property proxbid?

(b) In the opposite direction, W. Pollul [149] has proved the following theorem.

THEOREM 2.17. (i) *If T is a locally compact space, then $E = (C_0)_R(T)$ admits an equivalent norm $|||\cdot|||$ such that $(E, |||\cdot|||)$ does not have property proxbid, whenever E is infinite-dimensional.*

(ii) *If (T, ν) is a positive measure space, then $E = L_R^1(T, \nu)$ admits an equivalent norm $|||\cdot|||$ such that $(E, |||\cdot|||)$ does not have property proxbid, whenever E is infinite-dimensional.*

Both R. B. Holmes and B. R. Kripke [88] and W. Pollul [149] have also obtained as a by-product, that the space $E = c_0$ endowed with the equivalent norm

$$|||x||| = \sup_{1 \leqq n < \infty} |\xi_n| + \sum_{i=1}^{\infty} \frac{1}{2^i} |\xi_i|, \qquad x = \{\xi_n\} \in c_0, \tag{2.25}$$

does not have property proxbid; moreover [88], no $\Phi \in E^{**} \setminus \kappa(E)$ has an element of best approximation in $\kappa(E)$. In view of these results, J. Blatter [13] has raised a further problem.

Problem 2.5. If E is a nonreflexive Banach space, does E admit an equivalent norm $|||\cdot|||$ such that $(E, |||\cdot|||)$ fails to have property proxbid?

One can show that if E contains a complemented subspace G which admits an equivalent norm $|||\cdot|||$ such that $(G, |||\cdot|||)$ fails to have property proxbid, then E itself admits an equivalent norm for which it fails to have property proxbid. From this it follows, for example, that *for every nonreflexive subspace E of a Banach space with an unconditional basis* [173] *there exists an equivalent norm on E such that $(E, |||\cdot|||)$ does not have property proxbid.* Also, *if F is any infinite-dimensional Banach space with an unconditional basis, then* $\mathscr{C}(F, F)$ contains a complemented subspace isomorphic to c_0 (see, for example, [173, p. 492, Theorem 16.7]) and hence $E = \mathscr{C}(F, F)$ *admits an equivalent norm $|||\cdot|||$ such that $(E, |||\cdot|||)$ fails to have property proxbid.* A similar result can be also proved, with different methods, for a nonreflexive space E with a (not necessarily unconditional) basis, such that E^* has a basis.

(c) The examples of Banach spaces which fail the property proxbid, given in Theorem 2.17 and in the remarks made after Problem 2.5, have the common feature that they involve equipping naturally occurring Banach spaces with new equivalent "unnatural" norms. J. Blatter and G. L. Seever [16] have proved that *the disc algebra $E = A$, i.e., the space of continuous complex-valued functions on $D = \{\zeta \in C \mid |\zeta| \leqq 1\}$ which are analytic in* $\operatorname{Int} D = \{\zeta \in C \mid |\zeta| < 1\}$, *endowed with* *its natural norm* $\|x\| = \sup_{\zeta\in D}|x(\zeta)|$, *fails to have property proxbid.* This result suggests [16] the following two problems (see also Theorem 2.16 above).

Problem 2.6. Which uniform algebras E have the property proxbid?

Problem 2.7. Does every Banach space E admit an equivalent norm $|||\cdot|||$ such that $(E, |||\cdot|||)$ has property proxbid?

2.6. Transitivity of proximinality. Let us denote the statement "G is a proximinal linear subspace of the normed linear space E" by: $G \overset{(p)}{\subset} E$. We shall consider now the problem: to what extent is this relation transitive?

THEOREM 2.18. *Let E be a Banach space. The following statements are equivalent:*

1°. *For all Banach spaces F, G*

$$E \overset{(p)}{\subset} F, \quad F \overset{(p)}{\subset} G \quad \textit{imply} \quad E \overset{(p)}{\subset} G. \tag{2.26}$$

2°. *Same as* 1°, *with* $\dim F/E = \dim G/F = 1$.

3°. *For all Banach spaces F, G*

$$F \overset{(p)}{\subset} E, \quad E \overset{(p)}{\subset} G \quad \textit{imply} \quad F \overset{(p)}{\subset} G. \tag{2.27}$$

4°. *Same as* 2°, *with* $\dim E/F = \dim G/E = 1$.

5°. *E is reflexive.*

The implications 5° ⇒ 1° ⇒ 2° and 5° ⇒ 3° ⇒ 4° are immediate consequences of Corollary 2.1 and the fact that every closed linear subspace of a reflexive space is reflexive. The implications 2° ⇒ 5° and 4° ⇒ 5° have been proved recently by W. Pollul [149], using again James' theorem.

One can also consider a third type of transitivity property, with E "at the last place", but the following theorem of W. Pollul [149] shows that some nonreflexive Banach spaces also have this property.

THEOREM 2.19. *Let* $E = c_0$. *Then for all F, G,*

$$G \overset{(p)}{\subset} F, \quad F \overset{(p)}{\subset} E, \quad \dim F/G < \infty, \quad \dim E/F < \infty \quad \textit{imply} \quad G \overset{(p)}{\subset} E. \tag{2.28}$$

Naturally, by Corollary 2.1, every reflexive Banach space E also has property (2.28). On the other hand, there exist nonreflexive Banach spaces E which do not have property (2.28), even with $\dim F/G = \dim E/F = 1$, for example, $E = C_R(Q)$ (Q compact) and $E = L^1_R(T, \nu)$ ((T, ν) a positive measure space) whenever $\dim E = \infty$ (W. Pollul [149]).

Problem 2.8. (W. Pollul [149]). Which nonreflexive Banach spaces E have property (2.28) with $\dim F/G = \dim E/F = 1$?

Problem 2.9. (W. Pollul [149]). (i) Does $E = c_0$ have the property

$$G \overset{(p)}{\subset} F, \quad F \overset{(p)}{\subset} E \quad \text{imply} \quad G \overset{(p)}{\subset} E? \tag{2.29}$$

(ii) If not, then is every Banach space E with property (2.29) reflexive?

J. Blatter has observed [13, Remark 3.25] that $E = c_R$ and $E = l^\infty_R$ do not have property (2.29).

2.7. Proximinality and quotient spaces. The following theorem of Cheney and Wulbert [46] shows how proximinality is transmitted to and from quotient spaces.

THEOREM 2.20. *If G is a proximinal linear subspace of a normed linear space E and G_1 an arbitrary closed linear subspace of G, then G/G_1 is proximal in E/G_1.*

Conversely, if G is a closed linear subspace of E and G_1 a closed linear subspace of G such that G/G_1 is proximinal in E/G_1 (in particular, if G/G_1 is reflexive) and that G_1 is proximinal in E, then G is proximinal in E.

2.8. Very non-proximinal linear subspaces.

(a) DEFINITION 2.2. A set G in a metric space E is said to be *very non-proximinal* if G is closed and no element $x \in E \setminus G$ has an element of best approximation in G, i.e., if $\bar{G} = G$ and

$$\mathscr{P}_G(x) = \varnothing, \qquad x \in E \setminus G. \tag{2.30}$$

Some authors use for such subspaces the term *antiproximinal subspaces.*

Obviously, every very non-proximinal set is non-proximinal.

(b) We have the following simple characterization of very non-proximinal subspaces of normed linear spaces.

PROPOSITION 2.3. *For a linear subspace G of a normed linear space E the following statements are equivalent:*

1°. *G is very non-proximinal.*

2°. *We have*

$$\mathscr{P}_G^{-1}(0) = \{0\}. \tag{2.31}$$

3°. *We have* (where the sets $\mathscr{M}_f$ are defined by (2.14))

$$\mathscr{M}_f = \varnothing, \qquad f \in G^\perp \setminus \{0\}. \tag{2.32}$$

Indeed, if $z \in \mathscr{P}_G^{-1}(0) \setminus \{0\}$, then $z \in E \setminus G$ and $\mathscr{P}_G(z) \neq \varnothing$. Thus, $1° \Rightarrow 2°$. Furthermore, if $f \in G^\perp \setminus \{0\}$, $z \in E$, $\|z\| = 1$ and $f(z) = \|f\|$, then by § 1, Theorem 1.1, $z \in \mathscr{P}_G^{-1}(0) \setminus \{0\}$. Thus, $2° \Rightarrow 3°$. Finally, if $x \in E \setminus G$, $g_0 \in \mathscr{P}_G(x)$, then again by § 1, Theorem 1.1, there is an $f \in G^\perp$ with $\|f\| = 1$ such that $(x - g_0)/\|x - g_0\| \in \mathscr{M}_f$. Thus, $3° \Rightarrow 1°$, which completes the proof.

(c) From this proposition and from the observation made before Theorem 2.1 it follows that *a closed hyperplane G in E is very non-proximinal if and only if it is non-proximinal.* Consequently, by Theorem 2.14, *a Banach space E contains very non-proximinal linear subspaces if and only if E is nonreflexive*; moreover, in this case, for every $f \in E^*$ which has no maximal element, the closed hyperplane $G = \{x \in E \mid f(x) = 0\}$ is very non-proximinal. There arises [176] naturally the following problem.

Problem 2.10. Characterize those normed linear spaces (or Banach spaces) E which contain very non-proximinal subspaces G of codimension n, where $1 < n < \infty$.

Note that not every Banach space has this property. For example [176], if $E = B^*$, where B is a quasi-reflexive space of order 1 (i.e., $\dim B^{**}/\kappa(B) = 1$), then no subspace G of E with $1 < \operatorname{codim} G \leqq \infty$ is very non-proximinal; it is not known whether the converse of this is true, i.e., whether a Banach space E which contains no very non-proximinal subspace G with $1 < \operatorname{codim} G \leqq \infty$ must be the conjugate space of a quasi-reflexive space of order 1. At the other extreme, A. L. Garkavi has proved [68, Theorem 4] that *for every infinite compact space Q and every integer n with $1 \leqq n < \infty$, the space $E = C_R(Q)$ contains very non-proximinal subspaces of codimension n.*

E. W. Cheney and A. A. Goldstein [40] have observed that if G is the closed linear span in $E = C_R([a, b])$ of any infinite sequence $\{x_n\} \subset E$ such that (i) for each n, $\{x_1, \cdots, x_n\}$ is a Chebyshev system (see § 1) and (ii) $G \neq C_R([a, b])$ (for example, if $0 < a$, then the functions $x_n(t) = t^{2^n}$ satisfy (i) by Descartes' rule of signs and (ii) by the classical theorem of Müntz), then G is a very non-proximinal subspace of E, of infinite dimension and infinite codimension (this is an immediate consequence of the "alternation theorem" of § 1).

(d) We have seen in § 2.5 that the spaces $G = \mathscr{C}(H, H)$ (where H is a Hilbert space) and $G = c_0$ admit an equivalent norm such that if $E = (G, |||\cdot|||)$, then $\kappa(E)$ is very non-proximinal in E^{**}; also, any Hilbert space H admits an equivalent norm $|||\cdot|||$ such that if $F = (H, |||\cdot|||)$, then $\mathscr{C}(F, F)$ is very non-proximinal in $L(F, F)$. This suggests the following problems.

Problem 2.11. Does every nonreflexive Banach space G admit an equivalent norm $|||\cdot|||$ such that if $E = (G, |||\cdot|||)$, then $\kappa(E)$ is very non-proximinal in E^{**}?

Problem 2.12. Does every infinite-dimensional Banach space E admit an equivalent norm $|||\cdot|||$ such that if $F = (E, |||\cdot|||)$, then $\mathscr{C}(F, F)$ is very non-proximinal (or, at least, non-proximinal) in $L(F, F)$?

3. Uniqueness of elements of best approximation.

3.1. Characterizations of semi-Chebyshev and Chebyshev subspaces.

(a) The basic notions in connection with the uniqueness of elements of best approximation are given in

DEFINITION 3.1. A set G in metric space E is said to be

(i) a *semi-Chebyshev set*, if every element $x \in E$ has at most one element of best approximation in G, i.e., if

$$x \in E,\ g_1, g_2 \in \mathscr{P}_G(x) \ \text{ imply } \ g_1 = g_2; \tag{3.1}$$

(ii) a *Chebyshev set*, if it is simultaneously a proximinal and a semi-Chebyshev set, i.e., if every element $x \in E$ has exactly one element of best approximation in G.

Obviously, in Definition 3.1 the condition $x \in E$ can be replaced by: $x \in E \setminus G$.

The term "semi-Chebyshev" set has been proposed in the monograph [168]. The term "Chebyshev" set has been used previously by several authors, for example, by N. V. Efimov and S. B. Stechkin [59]. Some authors use for such sets the terms *U-set* and *EU-set* (or *Haar set*) respectively.

(b) Let us consider now the problem of characterization of semi-Chebyshev and Chebyshev (linear) subspaces G of a normed linear space E. From § 1, Theorem 1.14 we have that the following holds.

THEOREM 3.1. *A linear subspace G of a normed linear space E is a semi-Chebyshev subspace if and only if there do not exist $f \in E^*$, $x \in E$ and $g_0 \in G \setminus \{0\}$ such that*

$$\|f\| = 1, \tag{3.2}$$

$$f(g) = 0, \qquad g \in G, \tag{3.3}$$

$$f(x) = \|x\| = \|x - g_0\|. \tag{3.4}$$

Although this characterization of semi-Chebyshev subspaces is *not intrinsic* (i.e., it involves also elements of $E \setminus G$), it is convenient for applications, since one can deduce from it intrinsic characterizations of semi-Chebyshev subspaces in the usual concrete normed linear spaces, as we shall see below.

Let us also mention some characterizations of Chebyshev subspaces related to § 2, Proposition 2.1, which can be found in [46], [163] and [81] respectively.

PROPOSITION 3.1. *For a closed linear subspace G of a normed linear space E the following statements are equivalent:*

1°. *G is a Chebyshev subspace.*

2°. *We have*

$$E = G \oplus \mathscr{P}_G^{-1}(0), \tag{3.5}$$

where $\mathscr{P}_G^{-1}(0)$ is the set defined by § 1, *formula* (1.44) *and where $\oplus$ means that the sum decomposition of each element $x \in E$ is unique.*

3°. *G is proximinal and*

$$(\mathscr{P}_G^{-1}(0) - \mathscr{P}_G^{-1}(0)) \cap G = \{0\}. \tag{3.6}$$

4°. *G is proximinal and the restriction $\omega_G|_{\mathscr{P}_G^{-1}(0)}$ of the canonical mapping $\omega_G : E \to E/G$ to the set $\mathscr{P}_G^{-1}(0)$ is one-to-one.*

Indeed, the equivalence $1° \Leftrightarrow 2°$ is essentially proved by the same arguments as those used to prove § 2, Proposition 2.1, equivalence $1° \Leftrightarrow 2°$. Now, if 3° does not hold, then either G is non-proximinal, contradicting (3.5) (by § 2, Proposition 2.1) or (3.6) does not hold, say $y_1, y_2 \in \mathscr{P}_G^{-1}(0)$, $y_1 - y_2 \in G \setminus \{0\}$ and then $y_1 = 0 + y_1 = (y_1 - y_2) + y_2$, contradicting (3.5); thus, $2° \Rightarrow 3°$. Furthermore, if $y_1, y_2 \in \mathscr{P}_G^{-1}(0)$, $y_1 \neq y_2$, $\omega_G(y_1) = \omega_G(y_2)$, then $y_1 - y_2 \in (\mathscr{P}_G^{-1}(0) - \mathscr{P}_G^{-1}(0)) \cap G \setminus \{0\}$, contradicting (3.6); thus, $3° \Rightarrow 4°$. Finally, if $x \in E$, $g_1, g_2 \in \mathscr{P}_G(x)$, $g_1 \neq g_2$, then $x - g_1, x - g_2 \in \mathscr{P}_G^{-1}(0)$, $x - g_1 \neq x - g_2$, $\omega_G(x - g_1) = \omega_G(x - g_2)$ (because $x - g_1 - (x - g_2) = g_2 - g_1 \in G$), contradicting 4°; thus, $4° \Rightarrow 1°$, which completes the proof. Naturally, one obtains corresponding characterizations of semi-Chebyshev subspaces G by requiring instead of 2° that each element $x \in E$ have *at most* one sum decomposition and by omitting in 3° and 4° the condition of proximinality of G. It is also easy to show that *if G is a semi-Chebyshev subspace of E, then $\mathscr{P}_G^{-1}(0)$ is nowhere dense in E.*

Similarly to § 2, Theorem 2.2, one can prove the following (see, for example, [168, p. 105]).

THEOREM 3.2. *For a closed linear subspace G of a normed linear space E the following statements are equivalent:*

1°. *G is a semi-Chebyshev subspace.*

2°. *For every* $x \in E$ *there exists at most one element* $y \in E$ *satisfying* § 2, (2.4) *and* (2.5).

3°. *For every* $\Phi \in (G^{\perp})^*$ *there exists at most one element* $y \in E$ *satisfying* § 2, (2.6) *and* (2.7).

We recall that a linear subspace G of a normed linear space E is said to have *property* (U) if every functional $\varphi \in G^*$ has a unique extension with the same norm to the whole space E. We have the following relations of duality between Chebyshev subspaces and subspaces with property (U).

COROLLARY 3.1. *Let E be a normed linear space. Then*

(i) *A* $\sigma(E^*, E)$*-closed linear subspace* Γ *of* E^* *is a Chebyshev subspace if and only if* $\Gamma_{\perp} = \{x \in E | f(x) = 0 (f \in \Gamma)\} \subset E$ *has property* (U).

(ii) *If G is a proximinal linear subspace of E and* $G^{\perp} = \{f \in E^* | f(x) = 0 (x \in G)\} \subset E^*$ *has property* (U), *then G is a Chebyshev subspace of E.*

Indeed, (i) follows from Theorem 3.2 (see [168, pp. 107–109]) or, alternatively from Proposition 3.1, equivalence 1° ⇔ 4°, using § 2, formula (2.10) (for G, E replaced by Γ and E^* respectively) and the observation that $\omega_{\Gamma}: E^* — E^*/\Gamma$ carries each $f \in E^*$ into the set of all extensions of $f|_{\Gamma_{\perp}} \in (\Gamma_{\perp})^*$ to the whole space E (we have $\Gamma = (\Gamma_{\perp})^{\perp}$, since Γ is $\sigma(E^*, E)$-closed). More directly, part (i) also follows from § 1, Theorem 1.5 (i). Furthermore, if $G^{\perp} = (G^{\perp\perp})_{\perp} \subset E^*$ has property (U), then by part (i) $G^{\perp\perp}$ is a Chebyshev subspace in E^{**}, whence, since $\kappa(G) \subset G^{\perp\perp}$, (ii) follows. However, one can show that $E = c_0$ has finite-dimensional Chebyshev subspaces and that for no such subspace G has $G^{\perp}$ the property (U) (see [168, p. 108]), and thus the converse of (ii) is not true.

(c) In $E = C_R(Q)$ from Theorem 3.1 one obtains the following theorem of [164] which has been the first intrinsic characterization of semi-Chebyshev subspaces G of an arbitrary (finite or infinite) dimension of $E = C_R(Q)$ (see also [168, p. 117]).

THEOREM 3.3. *A linear subspace G of* $E = C_R(Q)$ *is a semi-Chebyshev subspace if and only if there do not exist a Radon measure* μ *on* Q, *two disjoint closed sets* $Y^+, Y^- \subset Q$ *and element* $g_0 \in G \setminus \{0\}$ *such that*

$$|\mu|(Q) = 1, \tag{3.7}$$

$$\int_Q g(q)\, d\mu(q) = 0, \qquad g \in G, \tag{3.8}$$

$$\mu \geqq 0 \text{ on } Y^+, \quad \mu \leqq 0 \text{ on } Y^- \quad \text{and} \quad Y^+ \cup Y^- \supset S(\mu), \tag{3.9}$$

$$g_0(q) = 0, \qquad q \in S(\mu), \tag{3.10}$$

where $S(\mu)$ *denotes the carrier of the measure* μ.

E. W. Cheney and D. E. Wulbert [46] have proved that *a linear subspace G of* $E = C_R(Q)$ *is a semi-Chebyshev subspace if and only if* 0 *is the only element of G which vanishes on an "*α_G*-set"* i.e., on a subset of Q of the form $\{q \in Q | \, |x(q)| = \|x\|\}$ for some $x \in \mathscr{P}_G^{-1}(0)$. Naturally, one can also deduce this easily from Theorem 3.1 or directly from Theorem 3.3.

Some other characterizations of semi-Chebyshev and Chebyshev subspaces of arbitrary dimension in the spaces $E = C(Q)$, $C_R(Q)$, $L^1(T, \nu)$, $L_R^1(T, \nu)$, $C^1(Q, \nu)$ and $C_R^1(Q, \nu)$ are given in [168, Chap. I, § 3]; we denote by $C^1(Q, \nu)$ (where Q is a compact space and ν a positive Radon measure on Q with carrier $S(\nu) = Q$) the linear subspace of $L^1(Q, \nu)$ consisting of the equivalence classes of the (complex or real) continuous functions on Q, endowed with the usual vector operations and with the norm $\|x\| = \int_Q |x(q)|\, d\nu(q)$. R. B. Holmes and B. R. Kripke [85] have given a necessary and sufficient condition in order that $C_R(T) \cap L_R^\infty(T, \nu)$ be a Chebyshev subspace of $L_R^\infty(T, \nu)$, where ν is a Borel measure on T with certain properties. Various conditions in order that $G = \{g_x(\cdot)|x \in \mathscr{D}\}$ be a semi-Chebyshev or a Chebyshev subspace of $C_F(Q)$ (see § 2.2(c), the part on best approximation by operator functions), which generalize Theorem 3.5 below, have been given, for example, by S. B. Stechkin [178], S. I. Zukhovitskii and G. I. Eskin [199] (see also the references in [168, p. 226]). E. W. Cheney and D. E. Wulbert [46] have proved that *a linear subspace G of $E = L_R^1(T, \nu)$ or $E = C_R^1(Q, \nu)$ is a semi-Chebyshev subspace if and only if* 0 *is the only element of G which vanishes on a "β_G-set", respectively, on a "γ_G-set"*, i.e., on a subset of T, respectively of Q, of the form $Z(x)$ for some $x \in \mathscr{P}_G^{-1}(0)$. Naturally, one can also deduce this result easily from Theorem 3.1 or directly from the characterizations of semi-Chebyshev subspaces of $L_R^1(T, \nu)$ and $C_R^1(Q, \nu)$, given in [168, p. 121 and p. 123 respectively].

(d) Let us consider now the particular case when $\dim G = n < \infty$. By § 2, Corollary 2.1, every such G is proximinal. Naturally, the preceding results on uniqueness of elements of best approximation are also valid in this particular case, but exploiting the assumption of finite dimensionality, we can obtain additional information. Thus, from Theorem 3.1 one obtains (see [168, pp. 210–211]):

THEOREM 3.4. *An n-dimensional linear subspace G of a normed linear space E is a Chebyshev subspace if and only if there do not exist h extremal points $f_1, \cdots, f_h$ of S_{E^*}, where $1 \leqq h \leqq n$ if the scalars are real and $1 \leqq h \leqq 2n - 1$ if the scalars are complex, h numbers $\lambda_1, \cdots, \lambda_h > 0$ with $\sum_{j=1}^h \lambda_j = 1$, and $x \in E$, $g_0 \in G \setminus \{0\}$, such that we have*

$$\sum_{j=1}^{h} \lambda_j f_j(g) = 0, \qquad g \in G, \tag{3.11}$$

$$f_j(x) = \|x\| = \|x - g_0\|, \qquad j = 1, \cdots, h. \tag{3.12}$$

The main difficulty in the proof of this result consists in establishing the bounds $h \leqq n$ and $h \leqq 2n - 1$ respectively (note that in § 1, Theorem 1.10 we had only the bounds $h \leqq n + 1$ and $h \leqq 2n + 1$ respectively).

(e) In particular, for $E = C(Q)$ (Q compact), we have the following classical theorem, given for real scalars by A. Haar [76] and for complex scalars by A. N. Kolmogorov [108].

THEOREM 3.5. *An n-dimensional linear subspace $G = [x_1, \cdots, x_n]$ of $E = C(Q)$ (Q compact) is a Chebyshev subspace if and only if $x_1, \cdots, x_n$ form a "Chebyshev system"* (i.e., every $\sum_{i=1}^n \alpha_i x_i \not\equiv 0$ has at most $n - 1$ zeros on Q).

This theorem follows as a simple particular case both from Theorem 3.3 for $\dim G = n < \infty$ (respectively, a complex version of it) and from Theorem 3.4 for $E = C(Q)$. These two proofs of Theorem 3.5, compared with the other function-theoretic proofs of Theorem 3.5 given previously by several authors, show clearly the advantages of applying the methods of functional analysis in the theory of best approximation.

Theorem 3.5 *remains valid also for the spaces* $E = I_A = \{x \in C(Q) | x(q) = 0 \ (q \in A)\}$ (where A is any closed subset of Q), endowed with the norm of $C(Q)$, replacing its condition by the condition that $x_1, \cdots, x_n$ form a Chebyshev system on $Q \setminus A$; hence, in particular, *Theorem* 3.5 *remains valid also in the spaces* $C_0(T)$, where T is locally compact (see [168, p. 218]).

We mention also the following characterization of finite-dimensional Chebyshev subspaces of $C_R(Q)$, due (for $Q = [a, b]$) to Y. Ikebe [93].

PROPOSITION 3.2. *A finite-dimensional subspace* G *of* $E = C_R(Q)$ (Q *compact*) *is a Chebyshev subspace if and only if*

$$\|g_0\| < 2\|x\|, \qquad x \in E \setminus \{0\},\ g_0 \in \mathscr{P}_G(x). \tag{3.13}$$

Some other characterizations of Chebyshev subspaces of finite dimension n in the spaces $E = C_F(Q)$ (Q compact, F a Banach space), $L^1(T, v)$, $L^1_R(T, v)$, $C^1(Q, v)$ and $C^1_R(Q, v)$ are given in [168, Chap. II, § 2]; some other references are given in [168, p. 237]. Let us mention here the following result of R. R. Phelps (see [168, p. 228, footnote]).

THEOREM 3.6. *An n-dimensional linear subspace* $G = [x_1, \cdots, x_n]$ *of* $E = L^1_R(T, v)$ (*where* $L^1_R(T, v)^* \equiv L^\infty_R(T, v)$) *is a Chebyshev subspace if and only if there does not exist a* $\beta \in L^\infty_R(T, v)$ *satisfying*

$$\int_T x_i(t)\beta(t)\, dv(t) = 0, \qquad i = 1, \cdots, n, \tag{3.14}$$

such that the set $\{t \in T | \, |\beta(t)| < \|\beta\|\}$ (*defined modulo a set of measure zero*) *be purely atomic and contain at most n atoms.*

We recall that an *atom* of T is a measurable set $A \subset T$ with $v(A) > 0$, such that if B is any measurable subset of A, then either $v(B) = 0$ or $v(A \setminus B) = 0$. In particular, from Theorem 3.6 it follows that *an n-dimensional linear subspace* G *of* $E = l^1_R$ *is a Chebyshev subspace if and only if for each* $f = \{\eta_n\} \in G^\perp \setminus \{0\}$ (*where we identify* $(l^1_R)^*$ *with* l^∞_R) *there are at least n indices* $i_1, \cdots, i_n$ *such that* $|\eta_{i_k}| < \|f\|, k = 1, \cdots, n$. Also, from Theorem 3.6 it follows [146] that *an n-dimensional linear subspace* Γ *of* $C_R(Q)^*$ (Q *compact*) *is a Chebyshev subspace if and only if for each bounded Baire measurable function* z *on* Q *satisfying* $|z(q)| \leqq 1$ ($q \in Q$) *and* $\int_Q z(q)\, d\mu(q) = 0$ ($\mu \in \Gamma$) *there are at least n points* $q_1, \cdots, q_n \in Q$ *such that* $|z(q_k)| < 1, k = 1, \cdots, n$.

Finally, we mention the following classical result, due to D. Jackson and M. G. Krein (see [168, p. 236]): *Let* $G = [x_1, \cdots, x_n]$ *be an n-dimensional subspace of* $E = C^1_R([0, 1], v)$, *where* v *is the Lebesgue measure, such that* $x_1, \cdots, x_n$ *form a Chebyshev system. Then* G *is a Chebyshev subspace.*

(f) Let us consider now closed linear subspaces G of finite codimension. From Theorem 3.1 one obtains the following result, due to A. L. Garkavi (see [168, p. 296]).

THEOREM 3.7. *A closed linear subspace G of codimension n of a normed linear space E, say $G = \{x \in E | f_1(x) = \cdots = f_n(x) = 0\}$ (with $f_1, \cdots, f_n \in E^*$ linearly independent), is a semi-Chebyshev subspace if and only if for every $f \in G^\perp \setminus \{0\}$ either the set $\mathcal{M}_f$ of all maximal elements of f is empty or $\mathcal{M}_f$ is of dimension $r = r(f) \leqq n - 1$ and contains $r + 1$ elements $x_0, x_1, \cdots, x_r$ such that*

$$\text{rank}\,(f_k(x_i))_{\substack{k=1,\cdots,n \\ i=0,1,\cdots,r}} = r + 1\,; \tag{3.15}$$

moreover, in this case for any $r + 1$ linearly independent elements $x_0, x_1, \cdots, x_r \in \mathcal{M}_f$ we have (3.15).

Since a subspace G is Chebyshev if and only if it is both semi-Chebyshev and proximinal, combining this characterization of semi-Chebyshev subspaces of finite codimension with either one of § 2, Theorems 2.1–2.4, *one obtains characterization of Chebyshev subspaces of finite codimension* (see, for example, [176, Theorem 3]); note that in this case for every $f \in G^\perp \setminus \{0\}$ we have $\mathcal{M}_f \neq \emptyset$ (e.g., by § 2, Theorem 2.2 or § 2, Theorem 2.4, implication 1° ⇒ 3°, considering the projection of A onto the first coordinate axis in H_n, i.e., the set $\{f_1(S_E)\}$).

(g) Applying Theorem 3.7 in the space $E = C_R(Q)$ (Q compact) and combining it with § 2, Theorem 2.5, one obtains the following result, due to A. L. Garkavi (see [168, pp. 315–320]).

THEOREM 3.8. *A closed linear subspace G of codimension n of $E = C_R(Q)$ (Q compact), say $G = \{x \in C_R(Q) | \mu_1(x) = \cdots = \mu_n(x) = 0\}$ (with $\mu_1, \cdots, \mu_n \in E^* \equiv \mathcal{M}_R^1(Q)$ linearly independent), is a Chebyshev subspace if and only if the following four conditions are satisfied:*

(i) *For every $\mu \in G^\perp \setminus \{0\}$ the space Q admits a Hahn decomposition with respect to μ into two closed sets Q^+ and Q^-.*

(ii) *Any two measures $\mu, \mu_0 \in G^\perp \setminus \{0\}$ are equivalent (i.e., each is absolutely continuous with respect to the other) on the set Q' of all limit points of Q.*

(iii) *For every $\mu \in G^\perp \setminus \{0\}$ the set $Q \setminus S(\mu)$ consists of at most $n - 1$ points.*

(iv) *For any r isolated points $q_1, \cdots, q_r$ of Q, where $1 \leqq r \leqq n - 1$, we have*

$$\text{rank}\,(\mu_k(\{q_i\}))_{\substack{k=1,\cdots,n \\ i=1,\cdots,r}} = r. \tag{3.16}$$

If Q contains at least n isolated points and G is proximinal, then condition (iii) *is necessary and sufficient in order that G be a Chebyshev subspace. If Q has no isolated point and G is proximinal then G is a Chebyshev subspace if and only if $S(\mu) = Q$ for all $\mu \in G^\perp \setminus \{0\}$.*

In the particular case *when Q is also metrizable*, A. L. Garkavi has proved [69] that *the above conditions hold if and only if every $\mu \in G^\perp \setminus \{0\}$ satisfies*: (i′) *For every convergent sequence $\{q_n\} \subset Q \setminus Q'$, the limit* $\lim_{n\to\infty} \text{sign}\, \mu(\{q_n\})$ *exists.* (ii′) $\text{Var}_{Q'}\, \mu = 0$. (iii′) *$\mu(\{q\}) = 0$ for at most $n - 1$ points $q \in Q \setminus Q'$. If $n > 1$ and Q is countable,* (ii′) *can be replaced by* : (ii″) *$\mu(\{q\}) = 0$ for all $q \in Q'$* [68].

Applying Theorem 3.7 in the space $E = L^1_R(T, \nu)$, where (T, ν) is a positive measure space such that $L^1_R(T, \nu)^* \equiv L^\infty_R(T, \nu)$, we obtain (see [168, p. 326] and [71, Theorem 8]; for an equivalent version, see also [131, Theorem 5]):

THEOREM 3.9. *A closed linear subspace G of codimension n of* $E = L^1_R(T, \nu)$ *(where* $L^1_R(T, \nu)^* \equiv L^\infty_R(T, \nu)$*), say* $G = \{x \in L^1_R(T, \nu) | \int_T x(t)\beta_i(t)\, d\nu(t) = 0\ (i = 1, \cdots, n)\}$ *(with* $\beta_1, \cdots, \beta_n \in L^\infty_R(T, \nu)$ *linearly independent), is a semi-Chebyshev subspace if and only if for each* $\beta \in G^\perp \setminus \{0\}$ *the set* $\operatorname{crit} \beta = \{t \in T | |\beta(t)| = \|\beta\|\}$ *is either of measure zero or consists of* $p \leqq n$ *atoms* $A_1, \cdots, A_p$ *for which*

$$\operatorname{rank}\,(\beta_k(A_j))_{\substack{k=1,\cdots,n\\ j=1,\cdots,p}} = p.$$

Combining this theorem with characterizations of proximinal subspaces, one obtains characterizations of Chebyshev subspaces of finite codimension in $L^1_R(T, \nu)$ (see, for example, [71, Theorem 9]).

In the space $E = C^1(Q, \nu)(S(\nu) = Q)$, E. W. Cheney and D. E. Wulbert [46] have proved that *a subspace G of codimension n such that* $\bigcup_{x \in \mathscr{P}_G^{-1}(0)} (T \setminus Z(x))$ *is finite and that each* $f \in G^\perp \setminus \{0\}$ *has a maximal element, is a semi Chebyshev subspace if and only if for each* $x \in \mathscr{P}_G^{-1}(0)$ *the set* $T \setminus Z(x)$ *contains at most n elements.*

Naturally, the foregoing results in general and in some of the above concrete normed linear spaces are applicable, in particular, also in the case of best approximation in conjugate spaces E^* by elements of $\sigma(E^*, E)$-closed linear subspaces Γ of finite codimension. However, using effectively the hypothesis that Γ is $\sigma(E^*, E)$-closed in E^*, one can obtain additional results (which present interest also because of the fact that they may be regarded, by Corollary 3.1 (i), as results on finite-dimensional subspaces $G \subset E$ with property (U)). For such *characterizations of* $\sigma(E^*, E)$*-closed Chebyshev subspaces* of finite codimension in general conjugate spaces E^* and in the concrete conjugate spaces $L^1_R(T, \nu)^*$, $C_R(Q)^*$, $L^\infty_R(T, \nu)^*$ and $((c_0)_R)^*$, see [168, Chap. III, § 3].

3.2. Existence of semi-Chebyshev and Chebyshev subspaces.

(a) It is obvious that every normed linear space E contains semi-Chebyshev linear subspaces; for example, every linear subspace G dense in E, with $G \neq E$, satisfies $\mathscr{P}_G(x) = \emptyset\ (x \in E \setminus G)$ and hence is a semi-Chebyshev subspace.

The problem of existence of semi-Chebyshev *closed* linear subspaces of Banach spaces is less trivial, but still has an affirmative answer, namely, *every Banach space has at least one semi-Chebyshev closed hyperplane.* Indeed, by § 2, every nonreflexive Banach space E contains very non-proximinal closed hyperplanes and any such closed hyperplane is clearly semi-Chebyshev. On the other hand, by a result of J. Lindenstrauss [124], in every reflexive Banach space E the unit cell S_E has an *exposed point*, i.e., a point $x \in \operatorname{Fr} S_E$ with the property that there exists a support hyperplane H of S_E such that $H \cap S_E = \{x\}$, and hence E obviously has a *Chebyshev hyperplane* (namely, $H - x$).

For Chebyshev subspaces of Banach spaces the situation is different, as shown by the following example of A. L. Garkavi (see [168, p. 114]): If I is a set of cardinality $> \mathfrak{c}$, then the (nonseparable) Banach space $E = E(I)$ of all bounded families $x = \{\xi_i\}_{i \in I}$ of scalars which have at most a countable number of nonzero coordinates ξ_i, endowed with the usual vector operations and with the norm $\|x\| = \sup_i |\xi_i|$, *has no Chebyshev subspace.* However, the following problem, raised essentially by A. L. Garkavi, in 1964 (see [168, p. 116] and [169]), is open.

Problem 3.1. Does there exist a *separable Banach* space E which has no Chebyshev subspace?

We have seen above that the assumption of separability of E in Problem 3.1 is essential. The assumption of completeness of E is also essential, since V. Klee and the present author have given (see [169]) the following example of a separable *noncomplete* normed linear space E which has no Chebyshev subspace: the dense linear subspace E of c_0 consisting of all almost-zero sequences (i.e., sequences with all coordinates $= 0$, except a finite number of them).

It is well known (see [168, p. 115]) that the space $E = c_0$ has no Chebyshev subspace of infinite dimension, but it has Chebyshev subspaces of any finite dimension. It is also known that $E = L_R^1([0, 1], \nu)$, where ν is the Lebesgue measure, has no Chebyshev subspace of finite dimension or of finite codimension, but still it has Chebyshev subspaces (see below). Therefore one might perhaps try to find a solution to Problem 3.1 by combining in some way the spaces c_0 and $L_R^1([0, 1], \nu)$. Furthermore, since in every separable conjugate Banach space $E = B^*$ the unit cell has at least one exposed point and hence E has a Chebyshev subspace, one may expect a positive solution of Problem 3.1 only among those separable Banach spaces which are not isometric to any conjugate Banach space (the spaces c_0 and $L_R^1([0, 1], \nu)$ do have this property; moreover, they are even not isomorphic to any conjugate Banach space).

(b) Now we shall consider the problem of existence of finite-dimensional Chebyshev subspaces G.

Problem 3.2 [176]. (i) Characterize those normed linear spaces (or Banach spaces) E which contain Chebyshev subspaces of dimension n, where $1 \leqq n < \dim E$. (ii) In particular, does every reflexive Banach space E contain such a subspace?

For finite-dimensional spaces E the answer to Problem 3.2 (ii) is affirmative, namely, we have

THEOREM 3.10. *Let* $2 \leqq \dim E = d < \infty$. *Then E contains a system* $G_1 \subset G_2 \subset \cdots \subset G_{d-1}$ *of Chebyshev subspaces such that* $\dim G_n = n, n = 1, \cdots, d - 1$.

Indeed, the existence of a one-dimensional Chebyshev subspace G_1 in every finite-dimensional Banach space E follows from a result of G. Ewald, D. G. Larman and C. A. Rogers [61] (this has solved in the affirmative the following equivalent geometric problem of V. Klee [104]: does every E with $2 \leqq \dim E = d < \infty$ possess a line G through the origin such that there exists no segment on Fr $S_E = \{x \in E | \, \|x\| = 1\}$ parallel to G?). Theorem 3.10 follows from this result by a simple induction as have observed B. Grünbaum and V. Klee (see [107]); indeed, one can find a one-dimensional Chebyshev subspace G_1 of E and a one-dimensional Chebyshev subspace G_2/G_1 of the $(d - 1)$-dimensional space E/G_1, where $G_1 \subset G_2 \subset E$, $\dim G_2 = 2$ and then by Theorem 3.18 below G_2 is a Chebyshev subspace of E, and the process can be continued. Moreover, this argument shows that in order to obtain an affirmative answer to Problem 3.2 (ii) it would be sufficient (and, obviously, also necessary) to prove that every reflexive Banach space contains *a one-dimensional Chebyshev subspace.*

Let us consider now the following problem: *What are* (i.e., characterize topologically) *the compact spaces Q for which $E = C(Q)$ has n-dimensional Chebyshev subspaces*? This problem presents interest also from the following general point of view: The classical theorem of Banach–Stone (according to which two compact spaces Q_1 and Q_2 are homeomorphic if and only if $C(Q_1)$ and $C(Q_2)$ are linearly isometric) shows that, theoretically, the metric-linear properties of the spaces $C(Q)$ are completely determined by the topological properties of the compact spaces Q and conversely; but, the effective, explicit study of this interdependence still presents many open problems and any nontrivial answer to the above question may be regarded also as a contribution to this study. Let us first observe that by the Haar–Kolmogorov theorem (Theorem 3.5 above) the problem is equivalent to the following: *What are the compact spaces Q which admit a real* (*or complex*) *Chebyshev system* $x_1, \cdots, x_n$ (naturally, we assume that Q consists of at least $n + 1$ points)?

For $n = 1$ the answer is obvious, since every compact space admits a Chebyshev system consisting of one function x_1, for example the function $x_1 \equiv 1$. For $n \geqq 2$ the answer is given by

THEOREM 3.11. *A compact space Q admits a real Chebyshev system $x_1, \cdots, x_n$ (or, what is equivalent: $C_R(Q)$ has an n-dimensional Chebyshev subspace) with $n \geqq 2$ if and only if Q is homeomorphic to a subset of the unit circumference $\{\{\xi_1, \xi_2\} | \xi_1^2 + \xi_2^2 = 1\}$. Moreover, if Q is homeomorphic to the whole unit circumference, then every real Chebyshev system on Q consists of an odd number of elements.*

This result has been conjectured by S. Mazur and proved by J. C. Mairhuber under the assumption that Q is a subset of a finite-dimensional Euclidean space; for general compact spaces it has been proved by K. Sieklucki and P. C. Curtis Jr. (see [168, pp. 218–222], where a proof due to I. J. Schoenberg and C. T. Yang is presented; for the last statement of Theorem 3.11 see, for example, [126, p. 26]).

For the case of complex scalars *the problem is still open* (only partial results are known, see [168, p. 222] and [141]).

From Theorems 3.11 and 3.5 one can deduce the following result, due to R. R. Phelps (see [168, p. 222]).

COROLLARY 3.2. *The space $E = L_R^\infty(T, \nu)$, where (T, ν) is a σ-finite positive measure space such that* $\dim E = \infty$, *has no Chebyshev subspace of finite dimension* $\geqq 2$ (however, it does have Chebyshev subspaces of dimension 1, even in the case of complex scalars).

In particular, this is valid for the space $E = l_R^\infty$, too.

One can also consider the analogous problem for the spaces $L^1(T, \nu)$. For real scalars we have the following result which is due, in the case when (T, ν) is σ-finite, to A. L. Garkavi (see [168, p. 233]).

THEOREM 3.12. *Let (T, ν) be a positive measure space such that $L_R^1(T, \nu)^* \equiv L_R^\infty(T, \nu)$ and let $n \geqq 1$. The space $E = L_R^1(T, \nu)$ has an n-dimensional Chebyshev subspace if and only if (T, ν) has at least n atoms.*

In particular, it follows that *if (T, ν) has no atoms* (for example, if $T = [0, 1]$ and ν is the Lebesgue measure), *then $L_R^1(T, \nu)$ has no Chebyshev subspaces of finite dimension*; this latter result is known to hold also for complex scalars (see [168, pp. 230–232]). On the other hand, from Theorem 3.12 it follows that $E = l_R^1$ does have Chebyshev subspaces of any finite dimension. Also, if $q_1, \cdots, q_n \in Q$ are distinct (Q compact), then $\{x \in C_R(Q) | x(q_k) = 0 (k = 1, \cdots, n)\}^{\perp}$ is an *n-dimensional Chebyshev subspace of* $E = C_R(Q)^*$ [146].

The analogous problem for the spaces $C^1(Q, v)$, i.e., the problem of characterization of the pairs (Q, v) (Q compact, $S(v) = Q$) for which $E = C^1(Q, v)$ has finite-dimensional Chebyshev subspaces, is still open. The theorem of D. Jackson and M. G. Krein mentioned in § 3.1(e) shows that, in contrast with the space $L_R^1([0, 1], v)$, where v is the Lebesgue measure, the space $C_R^1([0, 1], v)$ does have Chebyshev subspaces of any finite dimension.

(c) It is natural to consider the similar problems for subspaces of finite codimension.

Problem 3.3 [176]. (i) Does every normed linear space (or every Banach space) E contain a closed semi-Chebyshev subspace G of codimension n, where $1 < n < \dim E$? (ii) If the answer is negative, characterize those normed linear spaces (or Banach spaces) E which contain such subspaces G.

Obviously, every very non-proximinal subspace is a closed semi-Chebyshev subspace, whence, by § 2.8, *for every compact space Q and every integer n with* $1 \leqq n < \dim C_R(Q)$, *the space* $E = C_R(Q)$ *contains closed semi-Chebyshev subspaces G of codimension n.* Also, A. L. Garkavi has proved [71, Theorem 10] that *for every positive measure space (T, v) such that* $L_R^1(T, v)^* \equiv L_R^\infty(T, v)$ *and every integer n with* $1 \leqq n < \dim L_R^1(T, v)$, *the space* $E = L_R^1(T, v)$ *contains closed semi-Chebyshev subspaces G of codimension n.*

Problem 3.4 [176]. (i) Characterize those normed linear spaces (or Banach spaces) E which contain Chebyshev subspaces G of codimension n, where $1 < n < \dim E$. (ii) In particular, does every reflexive Banach space contain such a subspace?

For finite-dimensional spaces E the answer to Problem 3.4 (ii) is affirmative, as shown by Theorem 3.10 above.

For $n = 1$ the answers to Problems 3.4 (i) and (ii) are known, namely, we have seen at the beginning of this section that a normed linear space E contains a Chebyshev subspace of codimension 1 if and only if S_E has at least one exposed point (note that the existence of extremal points of S_E is not sufficient, since, for example, in $E = l^\infty([0, 1])$ there is no Chebyshev subspace of codimension n, where $1 \leqq n < \infty$, although $\mathscr{E}(S_E)$ is infinite; see [168, p. 301]) and every reflexive space has this property. One can give the following necessary (but not sufficient) condition [176]: *If a normed linear space E contains a Chebyshev subspace G of codimension n, then S_E has at least n linearly independent exposed points.*

Let us consider now the *problem of characterizing those compact spaces Q for which $C(Q)$ contains a Chebyshev subspace of finite codimension* $n \geqq 1$. The problem is solved for *metric* compact spaces Q and real scalars by the following result, due to A. L. Garkavi (see [168, p. 325, footnote], and [69]).

THEOREM 3.13. *For a compact metric space Q and any integer n with* $2 \leqq n < \infty$, *the space $C_R(Q)$ has a Chebyshev subspace G of codimension n if and only if*

$$Q = \overline{Q \setminus Q'} \tag{3.17}$$

(*i.e., Q coincides with the closure of the set of its isolated points*).

Hence, in particular, *the space* $E = c_R$ *has such subspaces.*

For nonmetrizable compact spaces Q only partial results are known, namely, conditions which are necessary (for example, that Q have at most a countable number of disjoint open subsets and that Q contain no open connected infinite subset) or which are sufficient in order that $C_R(Q)$ have Chebyshev subspaces of finite codimension $n \geqq 1$ or $n \geqq 2$ (see [168, Chap III]). However, for Stonian spaces Q, that is, for compact spaces which are extremally disconnected (i.e., such that the closure of every open set is open), the problem is solved by the following result of A. L. Garkavi [69, Theorem 5 and the subsequent remark].

THEOREM 3.14. *For a Stonian space Q and any integer n with $2 \leqq n < \infty$ the space $E = C_R(Q)$ has a Chebyshev subspace of codimension n if and only if there exist a Radon measure μ with $S(\mu) = Q$ and an $x \in C_R(Q)$ such that x is not constant on any subset of Q of positive measure which is not an isolated point of Q.*

In particular, from Theorem 3.14 it follows [71, Theorem 6] that *if (T, ν) is a σ-finite positive measure space for which there exists an $x \in L_R^\infty(T, \nu)$ such that x is not constant on any subset of T of positive measure which is not an atom, then $E = L_R^\infty(T, \nu)$ has Chebyshev subspaces of any finite codimension.* Hence, in particular, *the spaces $E = L_R^\infty([0, 1], \nu)$, where ν is the Lebesgue measure, and $E = l_R^\infty$, have such subspaces* [71].

However, as we have mentioned in part (a) of this section, *the space* $E = c_0$ *has no Chebyshev subspaces of infinite dimension* (in particular, none of finite codimension).

One can also consider the analogous problem for the spaces $L_R^1(T, \nu)$. The answer is given by the following result due, in the case when (T, ν) is σ-finite, to A. L. Garkavi (see [168, p. 331]), which shows that the condition for the existence of a Chebyshev subspace of codimension n coincides with that for the existence of an n-dimensional Chebyshev subspace (see Theorem 3.12 above).

THEOREM 3.15. *Let (T, ν) be a positive measure space such that $L_R^1(T, \nu)^* \equiv L_R^\infty(T, \nu)$ and let $n \geqq 1$. The space $E = L_R^1(T, \nu)$ has a Chebyshev subspace of codimension n if and only if (T, ν) has at least n atoms.*

In particular, it follows that *if (T, ν) has no atoms* (e.g., if $T = [0, 1]$ and ν is the Lebesgue measure), *then $L_R^1(T, \nu)$ has no Chebyshev subspaces of finite codimension.* On the other hand, $E = l_R^1$ does have Chebyshev subspaces of any finite codimension.

E. W. Cheney and D. E. Wulbert have proved [46, Theorem 34] that $E = C_R^1(Q, \nu)$ *(Q compact, $S(\nu) = Q$), contains a Chebyshev subspace of codimension n if and only if Q has at least n isolated points.*

(d) Finally, let us consider the similar problems for subspaces of infinite dimension and infinite codimension.

Problem 3.5. (i) Does every (infinite-dimensional) normed linear space (or every Banach space) E contain a closed semi-Chebyshev subspace G of infinite dimension and infinite codimension? (ii) If the answer is negative, characterize those normed linear spaces E which contain such subspaces.

We have seen in § 2.8, that the space $E = C_R([a, b])$ contains very non-proximinal, hence closed semi-Chebyshev subspaces of infinite dimension and infinite codimension. For the space $E = L_R^1([0, 1], \nu)$, where ν is the Lebesgue measure, we shall see below that the answer is affirmative, even with Chebyshev subspaces.

Problem 3.6. (i) Characterize those normed linear spaces (or Banach spaces) E which contain Chebyshev subspaces G of infinite dimension and infinite codimension. (ii) In particular, does every (infinite-dimensional) reflexive Banach space contain such a subspace?

Concerning concrete spaces, let us mention the following problem of A. L. Garkavi (see [169, p. 2] and [67, p. 96]).

Problem 3.7. Does the space $E = C_R([a, b])$ contain a Chebyshev subspace G of infinite dimension and infinite codimension?

D. E. Wulbert [193] has shown that there exist compact spaces Q such that the analogue of Problem 3.7 for $C_R(Q)$ has an affirmative answer. This suggests naturally to raise the problem of *characterizing those compact metric spaces Q for which $E = C(Q)$* (or $E = C_R(Q)$) *contains Chebyshev subspaces of infinite dimension and infinite codimension.* The answer is not known even for $E = c_R$ [132].

Recently L. Carleson and S. Jacobs have proved (see [37, Corollary of Theorem 1]) that in the metric of $L^\infty(\text{Fr } D, \nu)$, where $D = \{\zeta \mid |\zeta| \leqq 1\}$ and where ν is the Lebesgue measure, every $x \in C(\text{Fr } D)$ has a unique element of best approximation in the proximinal subspace H^∞ of $L^\infty(\text{Fr } D, \nu)$ (consisting of all $x \in L^\infty(\text{Fr D}, \nu)$ which are boundary values of bounded functions analytic in Int $D = \{\zeta \mid |\zeta| < 1\}$). It is natural to ask whether for each $x \in C(\text{Fr } D)$ this unique element of best approximation in H^∞ belongs already to $A = H^\infty \cap C(\text{Fr } D)$, since in this case A would be a Chebyshev subspace of $C(\text{Fr } D)$. However, the answer is negative, i.e., there exist $x \in C(\text{Fr } D)$ for which this unique element of best approximation in H^∞ does not belong to A ([37, Theorem 4]; as mentioned in [37], the first such example is due to V. M. Adamyan, D. Z. Arov and M. G. Krein).

The answer to the similar problems for the spaces $L_R^1(T, \nu)$ is known, namely, as has been observed by R. R. Phelps (see [168, p. 332]), it is easy to see that *if* (T, ν) *is a positive measure space and* $e \subset T$ *is a* ν*-measurable set with* $\nu(e) > 0$, $\nu(T \setminus e) > 0$, *then* $G = \{x \in L_R^1(T, \nu) | x(t) = 0$ ν*-a.e. on* $e\}$ *is a Chebyshev subspace of* $E = L_R^1(T, \nu)$, of infinite dimension and infinite codimension.

Also, J.-P. Kahane has proved [101] that $G_\Lambda = [e^{i\lambda t}]_{\lambda \in \Lambda}$, where $\Lambda \subset Z = \{\cdots, -2, -1, 0, 1, 2, \cdots\}$, is a Chebyshev subspace of $E = L^1(\text{Fr } D, \nu)$ (ν the Lebesgue measure) if and only if Λ is an infinite arithmetic progression with odd difference. Hence, in particular, $H^1 \equiv [e^{i\lambda t}]_{\lambda \in \{0,1,2,\cdots\}}$ and $H_0^1 \equiv [e^{i\lambda t}]_{\lambda \in \{1,2,\cdots\}}$ are Chebyshev subspaces of $E = L^1(\text{Fr } D, \nu)$.

3.3. Normed linear spaces in which all linear (respectively, all closed linear) subspaces are semi-Chebyshev (respectively, Chebyshev) subspaces. We recall that a normed linear space E is said to be *strictly convex* (or *rotund*) if the relations

$$x, y \in E \setminus \{0\}, \quad \|x + y\| = \|x\| + \|y\| \tag{3.18}$$

imply the existence of a $c > 0$ such that

$$y = cx. \tag{3.19}$$

It is well known and easy to show that this property is equivalent to each of the following properties: (i) Fr $S_E = \mathscr{E}(S_E)$; (ii) Fr S_E contains no segment; (iii) each $f \in E^* \setminus \{0\}$ has at most one maximal element.

Using Theorem 3.1, one obtains (see [168, p. 110])

THEOREM 3.16. *For a normed linear space E the following statements are equivalent:*

1°. *All linear subspaces of E are semi-Chebyshev subspaces.*

2°. *All linear subspaces of E of a certain fixed finite dimension n, where* $1 \leqq n < \dim E$, *are semi-Chebyshev (or, what is equivalent, Chebyshev) subspaces.*

3°. *All closed linear subspaces of E of a certain fixed finite codimension m, where* $1 \leqq m < \dim E$, *are semi-Chebyshev subspaces.*

4°. *E is strictly convex.*

Combining this with § 2, Theorem 2.14, we obtain the next theorem (see [168, p. 111]).

THEOREM 3.17. *For a Banach space E the following statements are equivalent:*

1°. *All closed linear subspaces of E are Chebyshev subspaces.*

2°. *All closed linear subspaces of E of a certain fixed finite codimension m, where* $1 \leqq m < \dim E$, *are Chebyshev subspaces.*

3°. *E is reflexive and strictly convex.*

In particular, since the spaces $E = L^p(T, \nu)$, $1 < p < \infty$, where (T, ν) is a positive measure space, and the Hilbert (=complete inner product) spaces $E = H$ satisfy 3°, it follows that *all closed linear subspaces of these spaces are Chebyshev subspaces.*

3.4. Semi-Chebyshev and Chebyshev subspaces and quotient spaces. The following results, corresponding to § 2, Theorem 2.20 have been proved by E. W. Cheney and D. E. Wulbert [46].

THEOREM 3.18. (i) *If G is a semi-Chebyshev linear subspace of a normed linear space E and* G_1 *a closed linear subspace of G, which is proximinal in E, then* G/G_1 *is a semi-Chebyshev subspace of* E/G_1.

(ii) *If G is a linear subspace of E and* $G_1 \subset G$ *a closed linear subspace of E such that* G/G_1 *is semi-Chebyshev in* E/G_1 *and that* G_1 *is semi-Chebyshev in E, then G is a semi-Chebyshev subspace of E.*

(iii) *If G is a closed linear subspace of E and* G_1 *is a closed linear subspace of G, such that* G/G_1 *is Chebyshev in* E/G_1 *and that* G_1 *is Chebyshev in E, then G is a Chebyshev subspace of E.*

3.5. Strongly unique elements of best approximation. Strongly Chebyshev subspaces. Interpolating subspaces.

(a) DEFINITION 3.2. Let G be a set in a metric space E. An element $g_0 \in G$ is said to be a *strongly unique element of best approximation* of an element $x \in E$ if there exists a constant $r = r(x, G)$ with $0 < r \leqq 1$, such that

$$\rho(x, g) \geqq \rho(x, g_0) + r\rho(g_0, g), \qquad g \in G. \tag{3.20}$$

In this case $\mathscr{P}_G(x) = \{g_0\}$, i.e., g_0 is the unique element of best approximation of x, since by $r > 0$ for every $g \in G \setminus \{g_0\}$ we have $\rho(x, g) > \rho(x, g_0)$. The following characterization of such elements is due, essentially (namely, for $g_0 = 0$ and $\|x\| = 1$), to D. E. Wulbert [195].

PROPOSITION 3.3. *Let G be a linear subspace of a real normed linear space E. An element $g_0 \in G$ is a strongly unique element of best approximation of an element $x \in E \setminus \bar{G}$ if and only if there exists a constant $r = r(x, G)$ with $0 < r \leqq 1$ such that*

$$\sup_{f \in M_{x-g_0}} f(g) \geqq r\|g\|, \qquad g \in G, \tag{3.21}$$

where

$$M_{x-g_0} = \{f \in E^* \mid \|f\| = 1, f(x - g_0) = \|x - g_0\|\}. \tag{3.22}$$

(b) DEFINITION 3.3. A Chebyshev set G in a metric space E is said to be a *strongly Chebyshev* set if every $x \in E$ has a strongly unique element of best approximation in G.

D. J. Newman and H. S. Shapiro (see [38, p. 80]) and, respectively, D. E. Wulbert [195] have proved

THEOREM 3.19. *In the spaces $C_R(Q)$ (Q compact) and $L_R^1(T, v)$ ((T, v) a positive measure space) every finite-dimensional Chebyshev subspace is a strongly Chebyshev subspace.*

On the other hand, D. E. Wulbert [195] has observed that *in a smooth normed linear space E no Chebyshev subspace is strongly Chebyshev.*

We recall that E is said to be smooth if for every $x \in E$ there exists only one $f = f_x \in E^*$ such that $\|f\| = 1, f(x) = \|x\|$.

(c) DEFINITION 3.4. An n-dimensional linear subspace G of a normed linear space E is called an *interpolating subspace* if for any n linearly independent extremal points $f_1, \cdots, f_n$ of S_{E^*} and any n numbers $c_1, \cdots, c_n$ there exists exactly one $g \in G$ such that

$$f_j(g) = c_j, \qquad j = 1, \cdots, n. \tag{3.23}$$

In arbitrary normed linear spaces such subspaces have been first considered in [162], where it was proved that *they are Chebyshev subspaces* (indeed, this is a consequence of Theorem 3.4; see [168, pp. 213–214]) and that the converse need not hold even if $\dim E < \infty$. Recently D. A. Ault, F. R. Deutsch, P. D. Morris and J. E. Olson [5] have studied best approximation by elements of interpolating subspaces, proving, among other results, the following.

THEOREM 3.20. *Every interpolating subspace G of a normed linear space E is a strongly Chebyshev subspace.*

From the Haar-Kolmogorov theorem (Theorem 3.5 above) it follows that *a finite-dimensional subspace G of $E = C(Q)$ is a Chebyshev subspace if and only if it is an interpolating subspace*; this, together with Theorem 3.20, implies again the first part of Theorem 3.19.

On the other hand, from Theorem 3.20 and the observation made after Theorem 3.19 it follows that a *smooth normed linear space E (in particular, the $L^p(T, v)$ spaces, for $1 < p < \infty$) contains no interpolating subspace.* For the L_R^1-spaces, D. A. Ault, F. R. Deutsch, P. D. Morris and J. E. Olson [5] have proved the following result, which should be compared with Theorem 3.12.

THEOREM 3.21. *For a σ-finite positive measure space (T, ν) the space $E = L_R^1(T, \nu)$ contains an interpolating subspace of dimension $n > 1$ if and only if T is the union of at least n atoms (or, equivalently, $L_R^1(T, \nu)$ is linearly isometric either to l_R^1 or to some $(l_m^1)_R$). Also, $L_R^1(T, \nu)$ contains a one-dimensional interpolating subspace if and only if T contains an atom.*

For complex scalars and $n > 1$ the situation is quite opposite, namely, *the complex spaces l^1 and l_m^1 contain no proper interpolating subspace of any finite dimension $n > 1$* (J. H. Biggs, F. R. Deutsch, R. E. Huff, P. D. Morris and J. E. Olson [8]) ; on the other hand, it is clear that the unit vector $\{1, 0, 0, \cdots\}$ in the complex spaces l^1 or l_m^1 spans a one-dimensional interpolating subspace.

3.6. Almost Chebyshev subspaces. k-semi-Chebyshev and k-Chebyshev subspaces. Pseudo-Chebyshev subspaces. We shall consider now some generalizations of semi-Chebyshev and Chebyshev subspaces.

(a) DEFINITION 3.5. A set G in a metric space E is called an *almost Chebyshev* set if the set of all $x \in E$ for which $\mathscr{P}_G(x)$ does not consist of a single element forms a set at most of the first category in E.

Almost Chebyshev linear subspaces of normed linear spaces have been introduced by A. L. Garkavi (see [168, p. 116]), since they have the advantage that *in every separable Banach space E there exist almost Chebyshev subspaces of any finite dimension.* However, the Banach space $E(I)$ of § 3.2 has no almost Chebyshev subspace and the space c_0 has no almost Chebyshev subspace of infinite dimension. For results on finite-dimensional almost Chebyshev subspaces of $C_R(Q)$ (Q compact) see [168, pp. 224–225].

(b) DEFINITION 3.6. A linear subspace G of a normed linear space E is called a *k-semi-Chebyshev subspace*, respectively a *k-Chebyshev subspace* (where k is an integer with $0 \leqq k < \infty$), if

$$-1 \leqq \dim \mathscr{P}_G(x) \leqq k, \qquad x \in E, \tag{3.24}$$

respectively if

$$0 \leqq \dim \mathscr{P}_G(x) \leqq k, \qquad x \in E. \tag{3.25}$$

We recall that $\mathscr{P}_G(x)$ is a convex set, since it is the intersection of the two convex sets G and $S(x, \rho(x, G))$ and that for a nonvoid convex set A in a linear space E the dimension $\dim A$ is defined as the dimension of the linear subspace of E spanned by $A - y$, where y is an arbitrary element of A; if $A = \emptyset$, then, by definition, $\dim A = -1$. Thus, the 0-semi-Chebyshev and 0-Chebyshev subspaces are nothing else than the usual semi-Chebyshev and, respectively, Chebyshev subspaces.

Most of the preceding results (e.g., Theorems 3.1, 3.3, 3.4, 3.5 and 3.16) admit extensions to k-semi-Chebyshev and k-Chebyshev subspaces (see [168, pp. 125–135 and 237–242]). We mention here that $-1 \leqq \dim \mathscr{M}_f \leqq k$ ($f \in G^\perp \setminus \{0\}$) is a sufficient (but not necessary) condition in order that G be a k-semi-Chebyshev subspace. Also, *all linear subspaces G of E are k-semi-Chebyshev subspaces if and only if E is "$(k + 1)$-strictly convex"*, i.e., for any $k + 2$ elements $x_0, x_1, \cdots, x_{k+1} \in E$ the equality $\|\sum_{i=0}^{k+1} x_i\| = \sum_{i=0}^{k+1} \|x_i\|$ implies the linear dependence of these elements. It is easy to show (see [168, pp. 128–130]) that this property is equivalent to each of the following properties: (i) Fr S_E contains no convex subset of dimension $> k$; (ii) $-1 \leqq \dim \mathcal{M}_f \leqq k (f \in E^* \setminus \{0\})$. However, only partial extensions of Theorem 3.11 are known, i.e., it is not known which are the compact spaces Q such that $C_R(Q)$ has k-Chebyshev subspaces of finite dimension n, where $n \geqq 2$ and $0 \leqq k \leqq n - 1$ (or, equivalently, such that Q admits systems $x_1, \cdots, x_n \in C_R(Q)$ "of Chebyshev rank k", i.e., with the property that there do not exist $n - k$ distinct points $q_1, \cdots, q_{n-k} \in Q$ and $k + 1$ linearly independent elements $g_0, g_1, \cdots, g_k \in G = [x_1, \cdots, x_n]$ such that $g_i(q_j) = 0$ for $j = 1, \cdots, n - k$ and $i = 0, \cdots, k$). For references on some partial results see [168, p. 242]; for more recent work see also [180], [53] (in $C(Q)$).

(c) DEFINITION 3.7. A linear subspace G of a normed linear space E is called a *pseudo-Chebyshev subspace* if

$$0 \leqq \dim \mathcal{P}_G(x) < \infty, \qquad x \in E. \tag{3.26}$$

Some authors use for these subspaces the term *EF-subspace*.

In particular, every finite-dimensional linear subspace and every k-Chebyshev subspace, $0 \leqq k < \infty$, is pseudo-Chebyshev. P. D. Morris has constructed examples of pseudo-Chebyshev subspaces of finite codimension of $E = l_R^\infty$ which are not Chebyshev subspaces [131].

One can show that for a proximinal subspace G of a normed linear space E, the condition $-1 \leqq \dim \mathcal{M}_f < \infty$ $(f \in G^\perp \setminus \{0\})$ is sufficient in order that G be a pseudo-Chebyshev subspace. Also, *all closed linear subspaces G of a normed linear space are pseudo-Chebyshev subspaces if and only if every convex subset of* Fr S_E *is finite-dimensional*; if E is a Banach space, another *equivalent condition* is that we have $0 \leqq \dim \mathcal{M}_f < \infty$ $(f \in E^* \setminus \{0\})$.

The following characterization of pseudo-Chebyshev subspaces of $E = C_R(Q)$, due to P. D. Morris [131], should be compared with Theorems 3.3 and 3.8 (although the latter is only for codim $G < \infty$).

THEOREM 3.22. *A proximinal linear subspace G of $E = C_R(Q)$ is a pseudo-Chebyshev subspace if and only if for every $\mu \in G^\perp \setminus \{0\}$ the set $Q \setminus S(\mu)$ is finite.*

3.7. Very non-Chebyshev subspaces. One can introduce the following notion, corresponding to § 2, Definition 2.2.

DEFINITION 3.8. A set G in a metric space E is said to be a *very non-Chebyshev* set if G is closed and if for no element $x \in E \setminus G$ does the set $\mathcal{P}_G(x)$ consist of a single element.

One can show (see, for example [168, pp. 114–116]) that all closed linear subspaces G of cardinality $> \mathfrak{c}$ of the space $E(I)$ of § 3.2 and all infinite-dimensional closed linear subspaces of $E = c_0$ are very non-Chebyshev subspaces.

4. Properties of metric projections.

4.1. The mappings π_G. Metric projections.

DEFINITION 4.1. (i) If G is a set in a metric space E, we shall denote by π_G any "selection" of the set-valued mapping $\mathcal{P}_G$, i.e., any one-valued mapping $\mathcal{D}(\pi_G) \to G$

(where $\mathscr{D}(\pi_G) = \{x \in E | \mathscr{P}_G(x) \neq \varnothing\}$) defined by

$$\pi_G(x) \in \mathscr{P}_G(x), \qquad x \in \mathscr{D}(\pi_G). \tag{4.1}$$

(ii) In the particular case when G is a Chebyshev set (hence $\mathscr{D}(\pi_G) = E$ and $\mathscr{P}_G(x)$ is the singleton $\{\pi_G(x)\}$ for each $x \in E$), π_G is called the *metric projection* of E onto G.

Some authors use for the metric projection π_G the term *normal projection*, or *best approximation operator*, or *nearest point map*, or *Chebyshev map*.

Some properties of the mappings π_G (and hence, in particular, of metric projections) onto linear subspaces of normed linear spaces are collected in the following theorem.

THEOREM 4.1. *Let E be a normed linear space and G a linear subspace of E. Then*

(i) *$G \subset \mathscr{D}(\pi_G)$ and $\pi_G(g) = g$ for all $g \in G$. Hence, if $x \in \mathscr{D}(\pi_G)$, then $\pi_G(x) \in \mathscr{D}(\pi_G)$ and we have*

$$\pi_G^2(x) = \pi_G(x), \qquad x \in \mathscr{D}(\pi_G), \tag{4.2}$$

i.e., the mapping π_G is idempotent.

(ii) *We have*

$$| \, \|x - \pi_G(x)\| - \|y - \pi_G(y)\| \, | \leqq \|x - y\|, \qquad x, y \in \mathscr{D}(\pi_G), \tag{4.3}$$

$$\|x - \pi_G(x)\| \leqq \|x\|, \qquad x \in \mathscr{D}(\pi_G), \tag{4.4}$$

$$\|\pi_G(x)\| \leqq 2\|x\|, \qquad x \in \mathscr{D}(\pi_G). \tag{4.5}$$

(iii) *π_G is continuous at every point $g_0 \in G$ (i.e., $x_n \in \mathscr{D}(\pi_G)$ and $x_n \to g_0 \in G$ imply that $\pi_G(x_n) \to \pi_G(g_0) = g_0$).*

(iv) *If G_1 is a linear subspace of G, we have*

$$\|x - \pi_G(x)\| \leqq \|x - \pi_{G_1}(x)\|, \qquad x \in \mathscr{D}(\pi_G) \cap \mathscr{D}(\pi_{G_1}). \tag{4.6}$$

If, in addition G is a semi-Chebyshev subspace, then

(v) *If $x \in \mathscr{D}(\pi_G)$ and $g \in G$, then $x + g \in \mathscr{D}(\pi_G)$ and we have*

$$\pi_G(x + g) = \pi_G(x) + \pi_G(g) = \pi_G(x) + g, \qquad x \in \mathscr{D}(\pi_G), g \in G, \tag{4.7}$$

i.e., π_G is quasi-additive.

(vi) *If $x \in \mathscr{D}(\pi_G)$ and α is an arbitrary scalar, then $\alpha x \in \mathscr{D}(\pi_G)$ and we have*

$$\pi_G(\alpha x) = \alpha \pi_G(x), \qquad x \in \mathscr{D}(\pi_G), \alpha = \text{scalar}, \tag{4.8}$$

i.e., π_G is homogeneous.

(vii) *If G is closed and $x_n \in \mathscr{D}(\pi_G)$, $\lim_{n \to \infty} x_n = x$, $\lim_{n \to \infty} \pi_G(x_n) = g$, then $x \in \mathscr{D}(\pi_G)$ and $\pi_G(x) = g$, i.e., π_G is closed.*

The proofs are straightforward (see [168, pp. 140–142 and 390]). Part (i) (and (vii)) shows that in the particular case when G is a Chebyshev subspace, the metric projection π_G is indeed a (nonlinear closed) projection of E onto G.

4.2. Continuity of metric projections. In the present section we shall be concerned with Chebyshev subspaces G for which the metric projection π_G is continuous. We know of no convenient short term for such subspaces in the literature (some authors use the term *EU*-subspace*) and we shall not introduce any special term for them.

(a) The first natural problem is that of characterization of those Chebyshev subspaces for which π_G is continuous. To this end, we recall that *if G is a Chebyshev subspace of a normed linear space E, then* by § 2, Proposition 2.1 and § 3, Proposition 3.1, $\omega_G|_{\pi_G^{-1}(0)}$ *is a (continuous) one-to-one mapping of* $\pi_G^{-1}(0)$ *onto* E/G. Moreover, in this case we have

$$(4.9) \qquad (\omega_G|_{\pi_G^{-1}(0)})^{-1}(x + G) = x - \pi_G(x), \qquad x + G \in E/G;$$

indeed, obviously $x - \pi_G(x) \in \pi_G^{-1}(0)$ and $\omega_G(x - \pi_G(x)) = x + G$. Now, the main characterization of Chebyshev subspaces G with continuous metric projection π_G is the following result, due to R. B. Holmes ([81, Theorem 6]; in the particular case when E/G is reflexive, this result also follows from [170, Theorem 3]).

THEOREM 4.2. *For a Chebyshev subspace G of a normed linear space E the metric projection* π_G *is continuous if and only if the restriction* $\omega = \omega_G|_{\pi_G^{-1}(0)}$ *of the canonical mapping* $\omega_G: E \to E/G$ *to the set* $\pi_G^{-1}(0)$ *is a homeomorphism of* $\pi_G^{-1}(0)$ *onto* E/G.

Indeed, observe that by (4.9) the diagram

(4.10)

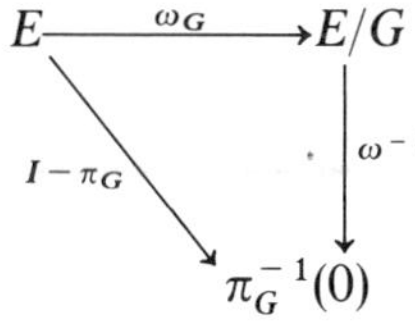

is commutative (where I denotes the identical mapping of E onto E). Therefore, if ω^{-1} is continuous, then $I - \pi_G = \omega^{-1}\omega_G$ is continuous, whence π_G is continuous. Conversely, if π_G is continuous, then so is $I - \pi_G$, whence, since (4.10) is commutative and since ω_G is an open mapping, it follows that ω^{-1} is continuous.

Since for every $x \in \pi_G^{-1}(0)$ we have $\omega^{-1}(x + G) = x - \pi_G(x) = x$, Theorem 4.2 can be also rephrased as follows: π_G *is continuous if and only if the relations* $x_n, x_0 \in \pi_G^{-1}(0)$, $\lim_{n\to\infty} \rho(x_n - x_0, G) = 0$ *imply* $\lim_{n\to\infty} x_n = x_0$. We shall see in the proofs of Corollary 4.1 and Theorems 4.5 and 4.7 below that Theorem 4.2 is useful for applications both in general and in concrete spaces.

COROLLARY 4.1. *Let E be a normed linear space. Then*

(i) *A* $\sigma(E^*, E)$*-closed Chebyshev subspace* Γ *of* E^* *admits a continuous metric projection* π_Γ *if and only if the (uniquely determined) extension map* $\chi: \varphi \in (\Gamma_\perp)^* \to f \in E^*$ *with* $f|_{\Gamma_\perp} = \varphi$, $\|f\| = \|\varphi\|$, *is continuous.*

(ii) *If G is a Chebyshev subspace of E, such that* $G^\perp \subset E^*$ *has property* (U) *and that the extension map* $\chi_1: \varphi \in (G^\perp)^* \to \Phi \in E^{**}$ *with* $\Phi|_{G^\perp} = \varphi$, $\|\Phi\| = \|\varphi\|$, *is continuous, then* π_G *is continuous.*

Indeed, part (i) follows from Theorem 4.2 observing that by the arguments of § 3, proof of Corollary 3.1, we have now the commutative diagram

(4.11)

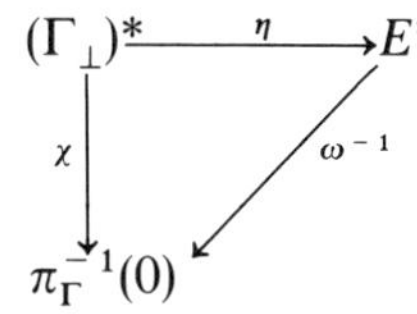

where η denotes the canonical linear isometry of $(\Gamma_\perp)^*$ onto $E^*/(\Gamma_\perp)^\perp = E^*/\Gamma$. One can also give a direct proof of part (i), observing that by § 1, formula (1.21), the diagram

(4.12)

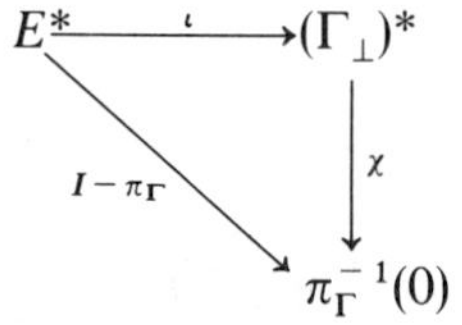

is commutative, where ι is the restriction map $f \to f|_{\Gamma_\perp}$ of E^* *onto* $(\Gamma_\perp)^*$; indeed, ι is an open mapping and thus one can apply the arguments of the above proof of Theorem 4.2. In order to prove part (ii) observe first that for any linear subspace G of E we have

$$\kappa(\mathscr{P}_G(x)) \subset \mathscr{P}_{G^{\perp\perp}}(\kappa(x)), \qquad x \in E; \tag{4.13}$$

indeed, this is contained in § 1, formula (1.25), but can be seen also directly, since

$$\begin{aligned} \|\kappa(x) - \kappa(g_0)\| &= \|x - g_0\| = \rho(x, G) = \sup_{\substack{f \in G^\perp \\ \|f\| \leqq 1}} |f(x)| \\ &= \sup_{\substack{f \in G^\perp \\ \|f\| \leqq 1}} |(\kappa(x))(f)| = \rho(\kappa(x), G^{\perp\perp}), \qquad x \in E,\ g_0 \in \mathscr{P}_G(x). \end{aligned}$$

Hence, if $G^{\perp\perp}$ is a Chebyshev subspace in E^{**} with continuous metric projection, then so is G in E. Now, since $G^\perp \subset E^*$ has property (U) and χ_1 is continuous, by § 3, Corollary 3.1 (i) and by (4.12) (with E^* and $\Gamma_\perp$ replaced by E^{**} and $G^\perp$ respectively) $G^{\perp\perp}$ is a Chebyshev subspace in E^{**} with $\pi_{G^{\perp\perp}}$ continuous, whence (ii) follows. Note that this argument amounts to the observation that under our assumptions $G^{\perp\perp}$ and G are Chebyshev subspaces and the diagram

(4.14)

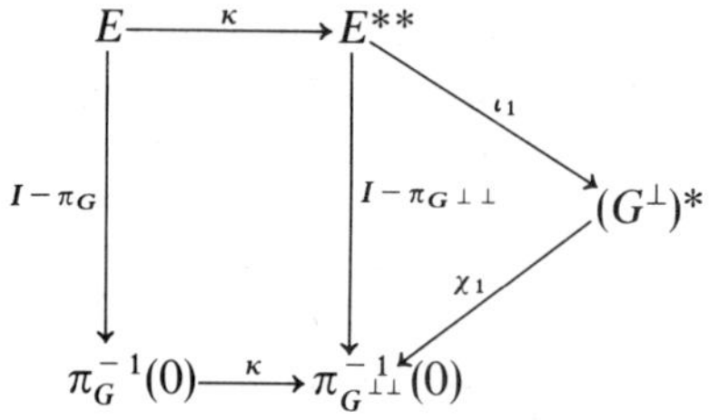

is commutative, where ι_1 is the restriction mapping. Part (i) of Corollary 4.1 is due to J. Lindenstrauss [122, § 7] and part (ii) has been given in [175, § 4].

Some other, more elementary, characterizations of the continuity of π_G due to E. W. Cheney–D. E. Wulbert [46] and R. B. Holmes [81] are collected in

PROPOSITION 4.1. *For a Chebyshev subspace G of a normed linear space E the following statements are equivalent*:

1°. *The metric projection π_G is continuous.*

2°. *π_G is continuous at each point of $\pi_G^{-1}(0)$.*

3°. *The direct sum decomposition* $E = G \oplus \pi_G^{-1}(0)$ *is topological (i.e.,* $\lim_{n\to\infty} x_n = x$ *if and only if* $\lim_{n\to\infty} \pi_G(x_n) = \pi_G(x)$ *and* $\lim_{n\to\infty}(x_n - \pi_G(x_n)) = x - \pi_G(x)$*).*

4°. $\pi_G|_{A_1(G)}$ *is continuous, where*

$$A_1(G) = \{x \in E \mid \|x - \pi_G(x)\| = 1\} = \{x \in E \mid \rho(x, G) = 1\}. \tag{4.15}$$

5°. *The functional*

$$\Phi_G(x) = \|\pi_G(x)\|, \qquad x \in E, \tag{4.16}$$

is continuous.

6°. *The mapping* $\psi_G : E \setminus G \to \pi_G^{-1}(0) \cap \operatorname{Fr} S_E$ *defined by*

$$\psi_G(x) = \frac{1}{\|x - \pi_G(x)\|}(x - \pi_G(x)), \qquad x \in E \setminus G, \tag{4.17}$$

is continuous.

Indeed, the implication 1° ⇒ 2° is obvious. Conversely, if 1° does not hold, say $x_n \to x_0, \pi_G(x_n) \not\to \pi_G(x_0)$, then $x_n - \pi_G(x_0) \to x_0 - \pi_G(x_0) \in \pi_G^{-1}(0)$, but $\pi_G(x_n - \pi_G(x_0)) = \pi_G(x_n) - \pi_G(x_0) \not\to 0$, contradicting 2°. Thus, 2° ⇒ 1°.

The equivalence 1° ⇔ 3° and the implication 1° ⇒ 5° are obvious. Conversely, if we have 5° and $x_n \to x_0 \in \pi_G^{-1}(0)$, then $\|\pi_G(x_n) - \pi_G(x_0)\| = \|\pi_G(x_n)\| \to \|\pi_G(x_0)\| = 0$, which proves that 5° ⇒ 2°.

The implication 1° ⇒ 6° is also obvious. Furthermore, if we have 6° and if x_n, $x_0 \in A_1(G)$, $x_n \to x_0$, then

$$\begin{aligned}\|\pi_G(x_n) - \pi_G(x_0)\| &\leqq \|x_n - \pi_G(x_n) - (x_0 - \pi_G(x_0))\| + \|x_n - x_0\| \\ &= \|\psi_G(x_n) - \psi_G(x_0)\| + \|x_n - x_0\| \to 0 \quad \text{as} \quad n \to \infty,\end{aligned}$$

which proves that 6° ⇒ 4°.

Assume, finally, that we have 4° and let $x_n \to x_0$. Since by Theorem 4.1 (iii) π_G is always continuous at the points $x_0 \in G$, we may assume that $x_0 \notin G$, hence $x_n \notin G$ for $n > N$. Then $x_n/\|x_n - \pi_G(x_n)\|$, $x_0/\|x_0 - \pi_G(x_0)\| \in A_1(G)$ and by Theorem 4.1 (ii), formula (4.3), $\|x_n - \pi_G(x_n)\| \to \|x_0 - \pi_G(x_0)\|$. Therefore, $x_n/\|x_n - \pi_G(x_n)\| \to x_0/\|x_0 - \pi_G(x_0)\|$, whence, by 4°,

$$\begin{aligned}\pi_G(x_n) &= \|x_n - \pi_G(x_n)\| \pi_G\left(\frac{x_n}{\|x_n - \pi_G(x_n)\|}\right) \\ &\to \|x_0 - \pi_G(x_0)\| \pi_G\left(\frac{x_0}{\|x_0 - \pi_G(x_0)\|}\right) = \pi_G(x_0);\end{aligned}$$

thus, 4° ⇒ 1°, which completes the proof of Proposition 4.1.

(b) No theorem is known in concrete spaces about characterization of Chebyshev subspaces G of arbitrary dimension with a continuous metric projection π_G.

(c) Let us consider now, in arbitrary normed linear spaces, the problem of characterization of Chebyshev subspaces G with continuous metric projection π_G, when we have restrictions on $\dim G$ or $\operatorname{codim} G$, or restrictions on the quotient space E/G.

THEOREM 4.3. *For every finite-dimensional Chebyshev subspace G of a normed linear space E, the metric projection π_G is continuous.*

The proof is straightforward, using a compactness argument (see [168, p. 251]).

For Chebyshev subspaces G of codimension 1, we have even a stronger property of π_G (see [168, pp. 142–145]):

THEOREM 4.4. *For every Chebyshev hyperplane G in a normed linear space E, the metric projection π_G is linear, and hence (e.g., by Theorem* 4.1 (ii) *or* (iii)) *continuous.*

We have the following characterizations of Chebyshev subspaces of finite codimension with continuous metric projections, due to Cheney–Wulbert [46] and Holmes [81] respectively.

THEOREM 4.5. *For a Chebyshev subspace G of finite codimension of a normed linear space E, the following statements are equivalent*:

1°. *π_G is continuous.*

2°. *$\pi_G^{-1}(0)$ is boundedly compact, that is, intersects every cell $S(x, r) \subset E$ in a conditionally compact set* (and hence, actually, in a compact set, since $\pi_G^{-1}(0)$ is closed).

3°. *$\pi_G^{-1}(0) \cap \operatorname{Fr} S_E$ is compact.*

Indeed, since $\dim E/G < \infty$, $S_{E/G}$ and $\operatorname{Fr} S_{E/G}$ are compact. Furthermore, by § 2, formula (2.10), $\pi_G^{-1}(0) \cap \operatorname{Fr} S_E = \{x \in E \mid \|\omega_G(x)\| = 1\}$, whence

$$\omega_G(\pi_G^{-1}(0) \cap \operatorname{Fr} S_E) = \operatorname{Fr} S_{E/G}, \tag{4.18}$$

$$\omega_G(\pi_G^{-1}(0) \cap S_E) = S_{E/G}, \tag{4.19}$$

where ω_G is the canonical map $E \to E/G$. Now, if π_G is continuous, then, by Theorem 4.2, $\omega = \omega_G|_{\pi_G^{-1}(0)}$ is a homeomorphism and hence, by (4.19), we have 2°. Finally, since the implication 2° $\Rightarrow$ 3° is obvious, let us assume that we have 3°. Then, by (4.18) and by the remarks made at the beginning of § 4.2(a), $\omega_G|_{\pi_G^{-1}(0) \cap \operatorname{Fr} S_E}$ is a one-to-one continuous mapping of the compact set $\pi_G^{-1}(0) \cap \operatorname{Fr} S_E$ onto the compact set $\operatorname{Fr} S_{E/G}$ and hence a homeomorphism. Therefore, since by (4.9) and (4.8) ω^{-1} is homogeneous (i.e., $\omega^{-1}(\alpha(x + G)) = \alpha\omega^{-1}(x + G)$), it follows that ω^{-1} is continuous on E/G and hence, by Theorem 4.2, π_G is continuous, which completes the proof of Theorem 4.5.

Naturally, one can also prove Theorem 4.5 directly, i.e., without using Theorem 4.2. Moreover, with a direct argument one can prove that the relations 3° $\Leftrightarrow$ 2° $\Rightarrow$ 1° remain valid for arbitrary Chebyshev subspaces G [46].

For strictly convex Banach spaces E we have also another useful characterization of Chebyshev subspaces G of finite codimension with continuous π_G, due to R. B. Holmes [81], in terms of the "spherical image map" $t:\mathscr{D}(t) \to E$, where $\mathscr{D}(t) \subset E^*$, defined as follows: for $f \in \mathscr{D}(t)$, $t(f) =$ the (unique) element $x \in E$ such that $f(x) = \|f\| \, \|x\|$, $\|x\| = \|f\|$, where $\mathscr{D}(t) = \{f \in E^* | t(f) \text{ exists}\}$.

THEOREM 4.6. *For a Chebyshev subspace G of finite codimension of a strictly convex Banach space E the metric projection π_G is continuous if and only if the restriction $t|_{G^\perp}$ of the spherical image map $t:\mathscr{D}(t) \to E$ to $G^\perp$ is continuous.*

Actually, Holmes has assumed [81] that E is reflexive, hence $\mathscr{D}(t) = E^*$, but in [51] it was observed that this assumption is superfluous, since $G^\perp \subset \mathscr{D}(t)$.

In connection with Theorem 4.6, note that we have always $t(G^\perp) = \pi_G^{-1}(0)$ and, if E is smooth, then also $t^{-1}(\pi_G^{-1}(0)) = G^\perp$. Furthermore, $t|_{G^\perp}$ is one-to-one if and only if E/G is smooth. Using the fact that t is a "duality map" and hence a "maximal monotone operator" in the sense of F. Browder (see, for example, [31]), R. B. Holmes [81] has proved that *the sufficiency part of Theorem* 4.6 *remains valid if instead of* codim $G < \infty$ *we assume only that the norm of E/G is Fréchet differentiable at every nonzero point.*

(d) We have the following characterization of Chebyshev subspaces of finite codimension with continuous metric projection in the spaces $C_R(Q)$, due to P. D. Morris [131].

THEOREM 4.7. *For a Chebyshev subspace G of finite codimension of $E = C_R(Q)$ (Q compact infinite), π_G is continuous (if and) only if G is a closed hyperplane.*

Indeed, since Q is an infinite compact space, there exists a limit point q_0 of Q. Since G is a Chebyshev subspace of finite codimension, by § 3, Theorem 3.8, condition (iii), for every $\mu \in G^\perp \setminus \{0\}$ the set $Q \setminus S(\mu)$ consists of isolated points, and thus $q_0 \in S(\mu)$. Hence, by § 1, Theorem 1.6, for every $x \in \pi_G^{-1}(0)$ we have $|x(q_0)| = \|x\|$. Consequently, for the set

$$A = \{x \in \pi_G^{-1}(0) \cap \mathrm{Fr}\, S_E | x(q_0) = 1\} \tag{4.20}$$

we have $A \neq \varnothing$ and $A \cup (-A) = \pi_G^{-1}(0) \cap \mathrm{Fr}\, S_E$. Also, it is obvious that $A \cap (-A) = \varnothing$ and that A and $-A$ are closed, whence $\pi_G^{-1}(0) \cap \mathrm{Fr}\, S_E$ is not connected. Now, if π_G is continuous, then by Theorem 4.2 and formula (4.18), $\omega_G|_{\pi_G^{-1}(0) \cap \mathrm{Fr}\, S_E}$ is a homeomorphism of $\pi_G^{-1}(0) \cap \mathrm{Fr}\, S_E$ onto $\mathrm{Fr}\, S_{E/G}$, whence $\mathrm{Fr}\, S_{E/G}$ is not connected, and hence we must have codim $G = \dim E/G = 1$. Conversely, if codim $G = 1$, then by Theorem 4.4 above π_G is continuous, which completes the proof. This argument shows that *Theorem* 4.7 *remains valid for every Chebyshev subspace G of $C_R(Q)$ such that $\bigcap_{\mu \in G^\perp \setminus \{0\},\ \mathscr{M}_\mu \neq \varnothing} S(\mu) \neq \varnothing$*; moreover, it also works *for every Chebyshev subspace G of a normed linear space E for which there exists an $f_0 \in E^*$ such that $|f_0(x)| = 1$ for all $x \in \pi_G^{-1}(0) \cap \mathrm{Fr}\, S_E$*. The original proof of Theorem 4.7, given in [131], was considerably more complicated; the present proof, based on the ideas of [121], is due to P. D. Morris [132] (see also [34, pp. 110–113] and [82, p. 169]).

A. Lazar, D. E. Wulbert and P. D. Morris have proved [121, Corollary 3.8] that *if G is a Chebyshev subspace of finite codimension $\geqq 3$ of $E = C_R(Q)$ (Q compact infinite), then every closed hyperplane containing G contains a point of discontinuity of π_G*. They have also given various extensions of Theorem 4.7 which imply, in particular, the following sharpening of Theorem 4.7 for the space $E = c_R$ [121, Corollary 3.10]: *For a Chebyshev subspace G of $E = c_R$, π_G is continuous (if and) only if either G is finite-dimensional or G is a hyperplane.* (However, as we have seen in § 3, the remarks made after Problem 3.7, the only infinite-dimensional Chebyshev subspaces of c_R that are known to exist are also of finite codimension).

P. D. Morris [131] has proved that *for a Chebyshev subspace G of finite codimension of $E = L_R^1(T, v)$, where (T, v) is a σ-finite positive measure space, π_G is continuous if and only if the set $\bigcup_{\beta \in G^\perp \setminus \{0\}}$* at $\{t \in T \mid |\beta(t)| = \|\beta\|\}$ *(where* at A *denotes the set of all atoms of A) is finite.*

E. W. Cheney and D. E. Wulbert have proved [46, Theorem 35] that *a proximinal subspace G of finite codimension n of $E = C_R^1(Q, \nu)$ (Q compact, $S(\nu) = Q$) is a Chebyshev subspace with continuous metric projection π_G if and only if $\bigcup_{x \in \mathscr{P}_G^{-1}(0)} (T \setminus Z(x))$ is finite and for each $x \in \mathscr{P}_G^{-1}(0)$ the set $T \setminus Z(x)$ contains at most n elements.*

Concerning subspaces of infinite dimension and infinite codimension, we have the following results of J. P. Kahane [101]: *For every Chebyshev subspace G of the space $E = L^1(\mathrm{Fr}\, D, \nu)$ (where* $\mathrm{Fr}\, D = \{\zeta \mid |\zeta| = 1\}$ *and ν is the Lebesgue measure), of the form $G_\Lambda = [e^{i\lambda t}]_{\lambda \in \Lambda}$, where $\Lambda \subset Z = \{\cdots, -2, -1, 0, 1, 2, \cdots\}$* (such subspaces have been characterized at the end of § 3.2), *the metric projection π_{G_Λ} is continuous.* Moreover, *π_{G_Λ} is uniformly continuous on the bounded subsets of $E = L^1(\mathrm{Fr}\, D, \nu)$ if and only if $\Lambda = (2p + 1)Z$, where p is a positive integer.*

(e) Let us consider now the problems of existence of continuous and discontinuous metric projections in normed linear spaces.

Theorems 4.3 and 4.4 show that in every normed linear space E there exist subspaces G for which π_G is continuous (namely, such is every finite-dimensional subspace G and every closed linear subspace G of codimension 1).

On the other hand, we have seen above various examples of existence of discontinuous metric projections. Thus, by Theorem 4.7, *for every Chebyshev subspace G of finite codimension $n \geqq 2$ of $E = C_R(Q)$ (Q compact infinite) π_G is discontinuous.* The abovementioned result in $L_R^1(T, \nu)$ implies that [131, Corollary 8] *if (T, ν) is σ-finite and* at *T is infinite, then there exists in $E = L_R^1(T, \nu)$ a Chebyshev subspace G of codimension* 2, *with discontinuous π_G*. The first example of a Chebyshev subspace G of a normed linear space E (namely, a subspace G of codimension 2 of $C_R([0, 1])^*$) with discontinuous metric projection π_G has been given in 1964, by J. Lindenstrauss [122], and then examples in other spaces by E. W. Cheney–D. E. Wulbert [46] (in l_R^1), R. B. Holmes–B. R. Kripke [87] and others. It has been conjectured in [87] that if G is a closed linear (hence Chebyshev) subspace of a strictly convex reflexive Banach space E, then π_G is continuous. However, this conjecture has been disproved, namely, B. R. Kripke (see [82, p. 169]) and independently, A. L. Brown [35] have constructed strictly convex Banach spaces E isomorphic to the separable Hilbert space l^2, which contain Chebyshev subspaces G with discontinuous π_G; in both examples the subspace G was of infinite codimension, but recently A. L. Brown [36] has also constructed *a strictly convex isomorph E of l^2 containing a subspace G of codimension* 2 *with discontinuous π_G.*

Now we shall show that the problem of existence of discontinuous metric projections may have a negative answer, i.e., there exist Banach spaces E in which for every Chebyshev subspace G the metric projection π_G is continuous and Banach spaces E such that all closed linear subspaces G are Chebyshev subspaces with continuous π_G. E. V. Ošman has proved [139, Theorem 3] that *a Banach space E has this latter property if and only if E is reflexive and the relations $\{x_n\}_0^\infty \subset E$, $\{f_n\}_0^\infty \subset E^*$, $f_n(x_n) = \|x_n\| = \|f_n\| = 1$, $n = 0, 1, 2, \cdots$, $\lim_{n \to \infty} \rho(x_n - x_0, [f_0, f_1, f_2, \cdots]_\perp) = 0$ imply $\lim_{n \to \infty} x_n = x_0$.*

Using Corollary 4.1 (i), J. Lindenstrauss [122] has proved that *if E^* is locally uniformly convex* (i.e., if the relations $f_n, f_0 \in \mathrm{Fr}\, S_{E^*}$, $\lim_{n \to \infty} \|f_n + f_0\| = 2$, imply

$\lim_{n\to\infty} \|f_n - f_0\| = 0$), *then every* $\sigma(E^*, E)$*-closed linear subspace* Γ *of* E^* *is a Chebyshev subspace with continuous* π_Γ. From this result the next proposition follows.

PROPOSITION 4.2. *In a uniformly convex Banach space E all closed linear subspaces G are Chebyshev subspaces with continuous* π_G.

We recall that a Banach space E is called *uniformly convex* if for every $\varepsilon > 0$ there is a $\delta(\varepsilon) > 0$ such that the relations $\|x\| = \|y\| = 1$, $\|x - y\| \geqq \varepsilon$ imply $\|x + y\| \leqq 2(1 - \delta(\varepsilon))$. It is well known (see, for example, [48]) that every uniformly convex space is reflexive and strictly convex.

On the other hand, it was proved in [166, Corollary 4], that *if E is a reflexive Banach space with "property* (H)" (i.e., such that the relations $x_n \to x_0$ weakly and $\lim_{n\to\infty} \|x_n\| = \|x_0\|$ imply $\lim_{n\to\infty} x_n = x_0$), *then the metric projection* π_G *onto any Chebyshev subspace G of E is continuous*; for spaces E which are, in addition, strictly convex, this was obtained by Ky Fan-I. Glicksberg [62]. This again implies Proposition 4.2, since it is well known that every uniformly convex space has property (H) (see, for example, [48]). J. M. Lambert has observed (see [82, p. 165]) that *there exists a strictly convex isomorph E of* l^2 *which does not have property* (H), *such that all* π_G *are continuous.*

We mention that for subspaces G of finite codimension (and even for G such that the norm of E/G is Fréchet differentiable at every nonzero point) Proposition 4.2 also follows from results of M. I. Kadec, who has proved [100, Lemma 2] that for a uniformly convex space E the spherical image map $t: E^* \to E$ is continuous and [100, Lemma 3] for codim $G < \infty$ the set $\pi_G^{-1}(0) \cap \operatorname{Fr} S_E$ is compact (and thus we can apply Theorem 4.6 or 4.5, combined with § 3, Theorem 3.17). Concerning metric projections in uniformly convex spaces, we also have the following stronger result of R. B. Holmes (see [82, p. 165]): *In a uniformly convex space E, the family of maps* $\{\pi_G | G$ *a closed linear subspace of* $E\}$ *is uniformly equicontinuous on any bounded subset of E.*

There also exist some results giving conditions in order that for all Chebyshev subspaces G of finite codimension the metric projection π_G be continuous. For example, R. B. Holmes [81] has observed that Theorem 4.6 leads to

PROPOSITION 4.3. *In a reflexive strictly convex Banach space E the metric projection onto every Chebyshev (or, equivalently, closed linear) subspace G of finite codimension is continuous if and only if the restriction of the spherical image map* $t: E^* \to E$ *to every finite-dimensional linear subspace of* E^* *is continuous* (or, in other words, if and only if t is continuous on E^* endowed with the "finite topology").

Independently, E. V. Ošman has proved [139, Theorem 4] that *in a reflexive strictly convex space E the metric projection onto every Chebyshev subspace G of finite codimension (respectively, of codimension* $\leqq k < \infty$*) is continuous if and only if the relations* $\{x_n\}_0^\infty \subset E$, $\{f_n\}_0^\infty \subset E^*$, $f_n(x_n) = \|x_n\| = \|f_n\| = 1$, $n = 0, 1, 2, \cdots$, $x_n \to x_0$ *weakly and* $\dim [f_n]_0^\infty < \infty$ *(respectively,* $\dim [f_n]_0^\infty \leqq k$*) imply* $\lim_{n\to\infty} x_n = x_0$.

In concrete spaces we have the following result of P. D. Morris [131, Corollary 7].

PROPOSITION 4.4. *If* (T, ν) *is a positive measure space such that* at *T is finite, then for every Chebyshev subspace G of finite codimension of* $E = L^1_R(T, \nu)$, *the metric projection* π_G *is continuous.*

Naturally, when applying the results on continuity of π_G (for example, Theorem 4.7 or Proposition 4.4), one must take into account the results of § 3 on characterization and existence of Chebyshev subspaces (in particular, of finite codimension, for example, Theorems 3.8, 3.13, 3.14).

(f) We conclude this section with the following theorem of Cheney and Wulbert [46] on continuity of metric projections in quotient spaces, corresponding to § 2, Theorem 2.20 and § 3, Theorem 3.18.

THEOREM 4.8. *Let G be a linear subspace of a normed linear space E and* G_1 *a subspace of G such that* G_1 *is a Chebyshev subspace of E with continuous* π_{G_1}. *Then G is a Chebyshev subspace of E with continuous* π_G *if and only if* G/G_1 *is a Chebyshev subspace of* E/G_1 *with continuous* π_{G/G_1}.

The converse of Theorem 4.8 is not valid: one can give [46] an example of $G_1 \subset G \subset E$, G_1 Chebyshev in E with $\operatorname{codim} G_1 = 2$, G Chebyshev in E with continuous π_G and G/G_1 Chebyshev in E/G_1 with continuous π_{G/G_1}, but π_{G_1} discontinuous.

4.3. Weak continuity of metric projections.

(a) *Weak sequential continuity of* π_G. The following analogues of some results of § 4.2 are due to R. B. Holmes [81].

THEOREM 4.9. (i) *If G is a Chebyshev subspace of a normed linear space E, such that* $\omega_G|_{\pi_G^{-1}(0)}$ *is a weak sequential homeomorphism of* $\pi_G^{-1}(0)$ *onto* E/G *(where* $\omega_G: E \to E/G$ *is the canonical mapping), then* π_G *is weakly sequentially continuous.*

(ii) *If G is a Chebyshev subspace of finite codimension and if* π_G *is continuous, then* π_G *is weakly sequentially continuous.*

(iii) *If G is any closed linear subspace of finite codimension of a normed linear space E, then* $\pi_G^{-1}(0)$ *is weakly sequentially closed.*

THEOREM 4.10. *For a Chebyshev subspace G of a reflexive Banach space E the following statements are equivalent:*

1°. π_G *is weakly sequentially continuous.*

2°. $\pi_G^{-1}(0)$ *is weakly sequentially closed.*

3°. $\omega_G|_{\pi_G^{-1}(0)}$ *is a weak sequential homeomorphism.*

Using the equivalence 1° ⇔ 2° above, J. Lambert [113] has shown (see [82, p. 171]) that *if G is a finite-dimensional subspace of* $E = L^p(T, \nu)$, *where* ν *is a separable nonatomic measure (e.g., the Lebesgue measure on* [0, 1]) *and where* $1 < p < \infty$, $p \neq 2$, *then* π_G *is not weakly sequentially continuous at any point of* $E = L^p(T, \nu)$.

COROLLARY 4.2. *For every Chebyshev subspace G of finite codimension of a reflexive Banach space E, the metric projection* π_G *is weakly sequentially continuous.*

Indeed, this follows immediately from Theorem 4.10, implication 2° ⇒ 1° and Theorem 4.9 (iii).

THEOREM 4.11. *If E is a reflexive strictly convex smooth Banach space such that mapping* $x \to \gamma_x$ of $E \setminus \{0\}$ *into* $\operatorname{Fr} S_{E^*}$ *defined by* $\gamma_x(y) = \lim_{t \to 0^+} (\|x + ty\| - \|x\|)/t$ *for all* $y \in E$, *is weakly sequentially continuous, then for every Chebyshev (or, equivalently, closed linear) subspace G of E the metric projection* π_G *is weakly sequentially continuous.*

Also, R. B. Holmes has shown that in the spaces $E = l^p$, where $1 < p < \infty$, the conclusion of Theorem 4.11 remains valid.

(b) *Weak continuity of* π_G. V. Klee [105] has given conditions on E which guarantee the weak continuity of π_G if $\dim G < \infty$, for example the following [105, Proposition 2.5]: *If for any pair of elements* $x, y \in E$ *the "equidistant set"*

$$P(x, y) = \{z \in E \mid \|x - z\| = \|y - z\|\} \tag{4.21}$$

is weakly closed, then for every finite-dimensional Chebyshev subspace G *of* E, π_G *is weakly continuous.* For the case when $\dim G = 1$, C. A. Kottman and Bor-Luh Lin [109] have given the following sharpening of this result: *If* G *is a Chebyshev subspace with* $\dim G = 1$ *and if for some* $g \in G$ *the set* $P(g, -g)$ *is weakly closed, then* π_G *is weakly continuous.* Let us also mention the following partial analogue of Theorem 4.10, due to Kottman and Lin [109] and Holmes [82, p. 170]: *For a finite-dimensional Chebyshev subspace* G *of a Banach space* E, π_G *is weakly continuous if and only if* $\pi_G^{-1}(0)$ *is weakly closed.* Using this result and the alternation theorem (§ 1, Theorem 1.13), R. B. Holmes has shown [82, p. 171] that *for the n-dimensional Chebyshev subspace* $G = [t^{k-1}]_{k=1}^n$ *of* $E = C_R([0, 1])$, π_G *is not weakly continuous.*

(c) For the *bw-topology* (see, for example, [55]), a result similar to Theorem 4.10, equivalence $1^\circ \Leftrightarrow 2^\circ$, has been given in [109, Theorem 3] and [82, p. 170].

An example of a one-dimensional Chebyshev subspace G of $E = c_0$ such that π_G is not weakly or bw-continuous, was given by Kottman and Lin [109].

4.4. Lipschitzian metric projections.

(a) *Pointwise Lipschitzian metric projections.* The following result is due, essentially, to G. Freud and E. W. Cheney (see [38, p. 82]).

PROPOSITION 4.5. *For every strongly Chebyshev subspace (hence, in particular, for every interpolating subspace)* G *of a normed linear space* E *the metric projection* π_G *is pointwise Lipschitzian, i.e., for each* $x \in E$ *there exists a constant* $\lambda = \lambda(G, x)$ *such that*

$$\|\pi_G(x) - \pi_G(y)\| \leqq \lambda \|x - y\|, \qquad y \in E. \tag{4.22}$$

Indeed, if $r = r(G, x)$ is as in § 3, formula (3.20), then, putting there $g_0 = \pi_G(x)$, $g = \pi_G(y)$, we obtain

$$\begin{aligned} r\|\pi_G(x) - \pi_G(y)\| &\leqq \|x - \pi_G(y)\| - \|x - \pi_G(x)\| \\ &\leqq \|x - y\| + \|y - \pi_G(y)\| - \|x - \pi_G(x)\| \\ &\leqq \|x - y\| + \|y - \pi_G(x)\| - \|x - \pi_G(x)\| \\ &\leqq \|x - y\| + \|y - x\| + \|x - \pi_G(x)\| - \|x - \pi_G(x)\| \\ &= 2\|x - y\|, \end{aligned}$$

and thus we may take $\lambda = 2/r$, which completes the proof.

The converse of Proposition 4.5 is not valid, since, for example, in $E = l^2$ we have (4.22) even with $\lambda = 1$, independent of G and x (since π_G is the orthogonal projection onto G), but $E = l^2$ has no strongly Chebyshev subspace G (by the remark made after § 3, Theorem 3.19, about smooth spaces).

Combining Proposition 4.5 with § 3, Theorem 3.19, it follows that *in the spaces* $E = C_R(Q)$ *and* $E = L_R^1(T, v)$ *for every finite-dimensional Chebyshev subspace* G *the metric projection* π_G *is pointwise Lipschitzian.*

On the other hand, R. B. Holmes and B. R. Kripke [87] have given an example of a Chebyshev line G in a 3-dimensional uniformly convex space E, such that π_G is not pointwise Lipschitzian.

(b) *Lipschitzian metric projections.* The main characterization of Chebyshev subspaces G with Lipschitzian metric projection π_G is the following analogue of Theorem 4.2, due to R. B. Holmes [81].

THEOREM 4.12. *For a Chebyshev subspace* G *of a normed linear space* E *the metric projection* π_G *is Lipschitzian if and only if* $\omega = \omega_G|_{\pi_G^{-1}(0)}$ *is a Lipschitzian homeomorphism of* $\pi_G^{-1}(0)$ *onto* E/G.

The proof is similar to that of Theorem 4.2 (the condition amounts to ω^{-1} being Lipschitzian). One can also give a corollary similar to Corollary 4.1.

Some other, more elementary, characterizations of Lipschitzian π_G, due to R. B. Holmes and B. R. Kripke [87], are collected in

PROPOSITION 4.6. *For a Chebyshev subspace* G *of a normed linear space* E, *the following statements are equivalent*:

1°. π_G *is Lipschitzian.*

2°. π_G *is uniformly continuous on* E.

3°. $\pi_G|_{A_1(\pi_G^{-1}(0))}$ *is bounded, where*

$$A_1(\pi_G^{-1}(0)) = \{x \in E \mid \rho(x, \pi_G^{-1}(0)) = 1\}. \tag{4.23}$$

4°. π_G *is "uniformly locally pointwise Lipschitzian", i.e., there exist two constants* $\lambda = \lambda(G) > 0$ *and* $\delta = \delta(G) > 0$ *such that the relations* $x \in \pi_G^{-1}(0) \cap \operatorname{Fr} S_E$, $\|x - y\| \leqq \delta$ *imply* $\|\pi_G(x) - \pi_G(y)\| \leqq \lambda\|x - y\|$.

R. B. Holmes and B. R. Kripke [87] have also observed that from a result of Lindenstrauss [123] one obtains, as a particular case, the following important necessary condition for π_G to be Lipschitzian.

THEOREM 4.13. *If the metric projection* π_G *onto a Chebyshev subspace* G *of a reflexive Banach space is Lipschitzian, then* G *is complemented in* E.

However, the condition in Theorem 4.13 is not sufficient. Indeed, R. B. Holmes and B. R. Kripke [87] have given even an example of a *one-dimensional* Chebyshev subspace G of l^p, where $2 < p < \infty$, such that π_G is not Lipschitzian.

(c) Combining Theorem 4.13 with a recent characterization of Banach spaces isomorphic to Hilbert spaces, due to J. Lindenstrauss and L. Tzafriri [125], we obtain

COROLLARY 4.3. *If for all closed linear subspaces* G *of a strictly convex reflexive Banach space* E, π_G *is Lipschitzian (or, in particular, uniformly Lipschitzian, i.e., with a constant* λ *independent of* G*), then* E *is isomorphic to a Hilbert space.*

R. B. Holmes and B. R. Kripke [87] have proved that *in the finite-dimensional* $L^p(T, v)$ *spaces, where* $2 < p < \infty$, π_G *is Lipschitzian for all Chebyshev (or equivalently, all linear) subspaces* G, *but not uniformly Lipschitzian, if* $\dim L^p(T, v) \geqq 3$. If E is a strictly convex space of dimension 2, then, by Theorem 4.4, π_G is uniformly Lipschitzian on E. R. B. Holmes and B. R. Kripke [87] have constructed examples of non-Hilbert spaces E of any finite dimension such that π_G is uniformly Lipschitzian on E.

In this connection we also have as follows (see [168, pp. 247–249 and 350]), where π_G is any selection of $\mathscr{P}_G$ (as in Definition 4.1 (i)).

THEOREM 4.14. *If E is a normed linear space of (finite or infinite) dimension $\geqq 3$ with the property that for every linear (not necessarily Chebyshev) subspace G of a certain fixed finite dimension n, or codimension n, where $1 \leqq n < \dim E$, we have $\|\pi_G(x)\| \leqq \|x\|$ for all $x \in \mathscr{D}(\pi_G)$ (hence, in particular, if π_G is "contractive" on $\mathscr{D}(\pi_G)$, i.e., Lipschitzian with constant 1), then E is linearly isometric to an inner product space.*

4.5. Differentiability of metric projections. The results of this section are due to R. B. Holmes and B. R. Kripke [87].

We recall that if G is a Chebyshev subspace of a normed linear space E and $x, y \in E$ and if the limit

$$\pi'_G(x, y) = \lim_{t \to 0} \frac{\|\pi_G(x + ty)\| - \|\pi_G(x)\|}{t} \tag{4.24}$$

exists, then $\pi'_G(x, y)$ is called the *Gâteaux derivative of π_G at x in the direction y.*

The following observations are immediate:

(i) $\pi'_G(x, cy) = c\pi'_G(x, y)$ if either side exists.

(ii) $\pi'_G(g, y) = \pi'_G(y)$ for all $g \in G$, $y \in E$.

(iii) If $x \in E \setminus G$ and either $\pi'_G(x, y)$ or $\pi'_G(\psi_G(x), y)$ exists, where ψ_G is as in Proposition 4.1, then both exist and are equal.

THEOREM 4.15. *If for a Chebyshev subspace G of a normed linear space E, $\pi'_G(x, y)$ exists for all $x \in \pi_G^{-1}(0) \cap \operatorname{Fr} S_E$, $y \in \operatorname{Fr} S_E$ and if*

$$\sup_{\substack{x \in \pi_G^{-1}(0) \cap \operatorname{Fr} S_E \\ y \in \operatorname{Fr} S_E}} \|\pi'_G(x, y)\| < \infty,$$

then π_G is Lipschitzian.

If $x \in E \setminus \{0\}$ and if for any $y, z \in E$ the function $N(s, t) = \|x + sy + tz\|$ is twice continuously differentiable in a neighbourhood of $(0, 0)$, then one can define a functional on $E \times E$ by

$$\langle y, z \rangle_x = \frac{\partial^2 N}{\partial s \, \partial t}(0, 0) \tag{4.25}$$

and one can show that $\langle y, z \rangle_x$ is a continuous symmetric bilinear form on $E \times E$, satisfying $\langle y, z \rangle_{cx} = (1/|c|)\langle y, z \rangle_x$ whenever $c \neq 0$. This form is used in the following theorem.

THEOREM 4.16. *Let $G = [x_1, \cdots, x_n]$ be an n-dimensional Chebyshev subspace of a normed linear space E and suppose that for some $x \in \pi_G^{-1}(0) \cap \operatorname{Fr} S_E$, $y \in \operatorname{Fr} S_E$, the function $f(s_1, \cdots, s_n, t) = \|x + ty - \sum_{k=1}^n s_k x_k\|$ is twice continuously differentiable in a neighborhood of the origin in R^{n+1}. Let π_x be the $n \times n$ matrix $(\pi_x)_{i,j} = \langle x_i, x_j \rangle_x$ and let $q_x(y)$ be the n-vector whose ith component is $\langle x_i, y \rangle_x$. Then,* *if π_x is invertible, $\pi'_G(x, y)$ exists and is given by*

$$\pi'_G(x, y) = \sum_{k=1}^{n} (\pi_x^{-1} q_x(y))_k x_k. \tag{4.26}$$

A straightforward computation shows that if $E = L_R^p(T,v), 2 < p < \infty, x \in \mathrm{Fr}\, S_E$, $y, z \in E$ and $\pi_{[y]}(x) = 0$, then $\langle y, z\rangle_x = (p-1)\int_T y(t)z(t)|x(t)|^{p-2}\, dv(t)$ and that the functional $x \to \langle y, z\rangle_x$ is continuous on E for any fixed $y, z \in E$. Moreover, using this observation and Theorem 4.16, one obtains

THEOREM 4.17. *For every finite-dimensional linear subspace G of $E = L_R^p(T, v)$, where $2 < p < \infty$, $\pi'_G(x, y)$ exists for all $x \in E \setminus G$, $y \in E$.*

The assumption here that $\dim G < \infty$, is essential, since one can give an example of an infinite-dimensional closed linear (hence Chebyshev) subspace G of l_R^p, where $1 < p < \infty$, $p \neq 2$, such that π_G is nondifferentiable and nonpointwise Lipschitzian.

We recall that if G is a Chebyshev subspace of a normed linear space E, π_G is said to be *Fréchet C^1 on the open set $E \setminus G$* if there exists a continuous mapping $u: E\setminus G \to L(E, G)$ (where $L(E, G)$ denotes the space of all continuous linear mappings of E into G, with the uniform norm), such that $\pi'_G(x, y)$ exists and

$$\pi'_G(x, y) = (u(x))(y), \qquad x \in E\setminus G, y \in E. \tag{4.27}$$

Using Theorem 4.15 one can show that *if $\dim E < \infty$ and if G is a Chebyshev subspace of E such that π_G is Fréchet C^1 on $E\setminus G$, then π_G is Lipschitzian.*

Some more results on the existence of $\pi'_G(x, y)$ whenever $\|x - x_0\| < \delta$ (for some $x \in E \setminus G$) and on π_G satisfying a Lipschitz condition for all x in a neighborhood of x_0 and all $y \in E$, are given in [87].

4.6. Linearity of metric projections.

(a) By Theorem 4.1 (vi), for any semi-Chebyshev subspace G the linearity of π_G on $\mathscr{D}(\pi_G)$ is equivalent to its additivity on $\mathscr{D}(\pi_G)$.

The main characterization of Chebyshev subspaces G with linear metric projection π_G is the following analogue of Theorem 4.2, due to R. B. Holmes [81].

THEOREM 4.18. *For a Chebyshev subspace G of a normed linear space E the metric projection π_G is linear if and only if $\omega = \omega_G|_{\pi_G^{-1}(0)}$ is an isometric (i.e., distance-preserving) mapping of $\pi_G^{-1}(0)$ onto E/G.*

Note that, as was observed before Theorem 4.2 and § 2, formula (2.10), for any Chebyshev subspace G, $\omega = \omega_G|_{\pi_G^{-1}(0)}$ is a one-to-one continuous norm-preserving mapping of $\pi_G^{-1}(0)$ onto E/G. One can also give a corollary of Theorem 4.18, similar to Corollary 4.1.

Some other characterizations of the linearity of π_G, given in [168, p. 144] and in [87, Theorem 3], are collected in

PROPOSITION 4.7. *For a semi-Chebyshev subspace G of a normed linear space E the following statements are equivalent:*

1°. *π_G is linear on $\mathscr{D}(\pi_G)$.*

2°. *$\pi_G^{-1}(0)$ is a closed linear subspace of E.*

3°. *$\pi_G^{-1}(0)$ is convex.*

If, in addition, G is proximinal (and hence a Chebyshev subspace), these statements are equivalent to the following:

4°. $\pi_G^{-1}(0)$ *contains a linear subspace F of E such that* $E = G + F$.

5°. *There exists a constant* K_G *such that*

$$\|\pi_G(x + y)\| \leqq K_G(\|\pi_G(x)\| + \|\pi_G(y)\|), \qquad x, y \in E. \tag{4.28}$$

6°. π_G *is continuously Gâteaux differentiable.*

In Theorem 4.4 it was established that for every Chebyshev hyperplane G in a normed linear space E, π_G is linear. There we also observed that whenever π_G onto a Chebyshev subspace G is linear, it is also continuous and hence a bounded linear projection; thus *a necessary condition in order that π_G be linear is that G be complemented in E.* Moreover, in this case, by Theorem 4.1 (ii) we have $1 \leqq \|\pi_G\| \leqq 2$ and $\|I - \pi_G\| = 1$. Some simple characterizations of the situation when $\|\pi_G\| = 1$, due to R. B. Holmes and B. R. Kripke [87], are collected in

PROPOSITION 4.8. *For a Chebyshev subspace G of a normed linear space E, such that π_G is linear, the following statements are equivalent:*

1°. $\|\pi_G\| = 1$.

2°. $(I - \pi_G)(x) \in \mathscr{P}_{\pi_G^{-1}(0)}(x)$ *for all* $x \in E$ (recall that $\pi_G^{-1}(0)$ is now a closed linear subspace, by Proposition 4.7).

If, in addition, $\pi_G^{-1}(0)$ is a Chebyshev subspace, these statements are equivalent to the following:

3°. $\pi_{\pi_G^{-1}(0)} = I - \pi_G$.

4°. $\pi_{\pi_G^{-1}(0)}^{-1}(0) = G$.

The following sufficient condition for the linearity of π_G, related to Theorem 4.6, was observed by R. B. Holmes [81] : *If for a Chebyshev (or, equivalently, closed linear) subspace G of a reflexive strictly convex Banach space E the restriction $t|_{G^\perp}$ of the spherical image map $t: E^* \to E$ to $G^\perp$ is linear, then π_G is linear.* However, the converse is not valid.

(b) There arises naturally the problem of characterizing in the usual concrete normed linear spaces E the Chebyshev subspaces G for which π_G is linear. In this direction we have the following result of P. D. Morris [131, Theorem 9].

THEOREM 4.19. *For a Chebyshev subspace G of finite codimension of* $E = L_R^1(T, \nu)$, *where (T, ν) is a positive measure space, π_G is linear if and only if there exists a* $\beta_0 \in G^\perp \setminus \{0\}$ *such that*

$$\text{at}\,\{t \in T \mid |\beta_0(t)| = \|\beta_0\|\} = \bigcup_{\beta \in G^\perp \setminus \{0\}} \text{at}\,\{t \in T \mid |\beta(t)| = \|\beta\|\}. \tag{4.29}$$

An example of a Chebyshev subspace G of codimension 2 of $E = l_R^1$ with π_G continuous but not linear, has been given by P. D. Morris [131]. T. Ando [2] has proved (see [82, p. 163]) that *for a closed linear subspace G* (hence Chebyshev, with continuous metric projection π_G) *of* $E = L^p(T, \nu)$, *where ν is a finite positive measure and $1 < p < \infty$, π_G is linear if and only if the quotient space E/G is linearly isometric to some other $L^p(T_1, \nu_1)$ space.* E. W. Cheney and K. H. Price have proved (see [44] or [82, p. 163]) that *for every finite-dimensional Chebyshev subspace G of*

$E = C_R(Q)$, *where* Q *is a perfect compact space (i.e., having no isolated points),* π_G *is nonlinear*. More generally, by the same proof and by [64], this result *remains valid for the space* $E = C_F(Q)$, *where* Q *is a perfect compact space and* F *a Banach space*. Indeed, in [64] it was proved that if u is any compact linear operator on such a space E, then $\|I + u\| = 1 + \|u\|$. Hence, if π_G were linear, where $\dim G < \infty$, then we would have $\|I - \pi_G\| = 1 + \|\pi_G\| \geqq 2$, in contradiction with Theorem 4.1 (ii), formula (4.4), which completes the proof.

(c) The following result (see [168, pp. 249 and 351] and [170]), somewhat related to Corollary 4.2 and Theorem 4.14, gives, in particular, a characterization of the spaces E in which all π_G are linear.

THEOREM 4.20. *If E is a normed linear space of (finite or infinite) dimension* $\geqq 3$, *with the property that every closed linear subspace G of a given fixed finite dimension n, or codimension n, where* $1 \leqq n \leqq \dim E - 2$, *respectively where* $2 \leqq n \leqq \dim E - 1$, *is a Chebyshev subspace such that the metric projection* π_G *is linear, then E is linearly isometric to an inner product space.*

For codim $G \geqq 3$ one can prove more, namely the following (see [168, p. 352] and [170]).

THEOREM 4.21. *If E is a normed linear space of (finite or infinite) dimension* $\geqq 4$, *such that every closed linear subspace G of a certain fixed finite codimension n, where* $3 \leqq n \leqq \dim E - 1$, *is a semi-Chebyshev subspace with the mapping* π_G *satisfying*

$$\pi_G(x + y) = \pi_G(x) + \pi_G(y), \qquad x, y, x + y \in \mathscr{D}(\pi_G), \tag{4.30}$$

then E is linearly isometric to an inner product space.

However [170], it is not known what happens if the same conditions are satisfied for every closed linear subspace G of codimension 2.

Finally, let us mention that in some cases for a certain (increasing or decreasing) sequence $\{G_n\}$ of Chebyshev subspaces of a space E, each π_{G_n} is linear, but not for all Chebyshev subspaces G of E is π_G linear. For example, in $E = l^1$, the increasing sequence $G_n = [e_1, \cdots, e_n]$ (where $e_k = \{\underbrace{0, \cdots, 0}_{k-1}, 1, 0, \cdots\}$) and the decreasing sequence $G_n^{(1)} = \{x = \{\xi_k\} \in E | \xi_1 = \cdots = \xi_n = 0\}$, $n = 1, 2, \cdots$, have this property; in this example we also have $\dim G_n = \operatorname{codim} G_n^{(1)} = n$ for all $n = 1, 2, \cdots$. In the general case, this property is related to the existence of a "Schauder basis" in E.

4.7. Semi-continuity and continuity of set-valued metric projections.

(a) In the foregoing we have considered mainly the important particular case of metric projections onto Chebyshev subspaces. Now we shall consider the metric projection in its full generality, namely, as a set-valued mapping.

DEFINITION 4.2. For a set G in a metric space E, the mapping $\mathscr{P}_G : x \to \mathscr{P}_G(x)$ of E into 2^G (=the collection of all subsets of G) is called the *set-valued metric projection* of E onto G.

Thus, in contrast with our convention for π_G (see Definition 4.1 (i)), in this more general setting the domain of $\mathscr{P}_G$ is not only the set $\mathscr{D}(\pi_G) = \{x \in E | \mathscr{P}_G(x) \neq \emptyset\}$, but the whole space E. Let us also note that if $G \subset E$ is closed, then $\mathscr{P}_G(x)$ is closed for each $x \in E$.

Some of the foregoing results on properties of the mappings π_G admit natural extensions to the set-valued case. For example, Theorem 4.1 (v), (vi) extends to

$$\mathscr{P}_G(x + g) = \mathscr{P}_G(x) + \mathscr{P}_G(g) = \mathscr{P}_G(x) + g, \qquad x \in \mathscr{D}(\pi_G), g \in G, \tag{4.31}$$

$$\mathscr{P}_G(\alpha x) = \alpha \mathscr{P}_G(x), \qquad x \in \mathscr{D}(\pi_G), \quad \alpha = \text{scalar}. \tag{4.32}$$

Indeed, by § 1, formula (1.45) and since $\alpha \mathscr{P}_G^{-1}(0) = \mathscr{P}_G^{-1}(0)$ (§ 1, Proposition 1.1), we have

$$\begin{aligned}\mathscr{P}_G(\alpha x + g) &= (\alpha x + g + \mathscr{P}_G^{-1}(0)) \cap G \\ &= (\alpha x + \mathscr{P}_G^{-1}(0)) \cap G + g = \alpha(x + \mathscr{P}_G^{-1}(0)) \cap G + g \\ &= \alpha \mathscr{P}_G(x) + g,\end{aligned}$$

which proves (4.31) and (4.32).

In the present section we shall consider the semi-continuity and continuity properties of set-valued metric projections. Since for set-valued mappings there exist several notions of upper and lower semi-continuity, the first natural problem is the following: Which ones of these notions (or which new notions, to be introduced) are suitable for the study of metric projections? We shall consider below several such notions. Let us recall the following convention on $\varnothing$, which will be used in the sequel: $\rho(x, \varnothing) = \infty$ $(x \in E)$.

(b) Semi-continuity and continuity properties of set-valued metric projections in normed linear spaces and, more generally, in metric spaces, have been first investigated by K. Tatarkiewicz [181], who has studied semi-continuity in the sense of C. Kuratowski. We recall that a mapping $\mathscr{U}: E \to 2^G$ (where E, G are metric spaces) is said to be *upper Kuratowski semi-continuous* (u.K.s.c.) *at* x_0, respectively *lower Kuratowski semi-continuous* (l.K.s.c.) *at* x_0, if the relations $\lim_{n\to\infty} x_n = x_0$, $y_n \in \mathscr{U}(x_n)$, $n = 1, 2, \cdots$, $\lim_{n\to\infty} y_n = y_0$ imply $y_0 \in \mathscr{U}(x_0)$, respectively if the relations $\lim_{n\to\infty} x_n = x_0$, $y_0 \in \mathscr{U}(x_0)$ imply $\lim_{n\to\infty} \rho(y_0, \mathscr{U}(x_n)) = 0$; $\mathscr{U}$ is called u.K.s.c., respectively l.K.s.c., if it is u.K.s.c., respectively l.K.s.c., at each $x_0 \in E$. In the particular case when $\mathscr{U}(x)$ is a singleton for each $x \in E$, these notions reduce to the usual closedness, respectively continuity, of $\mathscr{U}$. In [166] it was proved that if G is any proximinal set in a metric space E, then $\mathscr{P}_G$ is u.K.s.c.; this, together with the conveniently extended Theorem 4.1 (vii), shows that *for any set G in a metric space E, $\mathscr{P}_G$ is u.K.s.c.* Therefore, the notion of upper Kuratowski semi-continuity is not suitable for the study of metric projections, being too weak. We shall see below that for lower Kuratowski semi-continuity the situation is different, since it is equivalent to the usual lower semi-continuity; actually, already K. Tatarkiewicz [181] has given an example that the metric projection $\mathscr{P}_G$ onto a finite-dimensional subspace of a normed linear space E need not be l.K.s.c. Let us also mention that $\mathscr{U}: E \to 2^G$ is said to be Kuratowski *continuous at* x_0, if it is both u.K.s.c. and l.K.s.c. at x_0.

(c) We recall that if E, G are two metric spaces and $x_0 \in E$, a mapping $\mathscr{U}$ of E into 2^G is called:

(i) *upper semi-continuous* (u.s.c.) *at* x_0, respectively *lower semi-continuous* (l.s.c.) *at* x_0, if for every open set $M \subset G$ such that $\mathscr{U}(x_0) \subset M$, respectively such that $\mathscr{U}(x_0) \cap M \neq \emptyset$, there exists in E an open neighborhood V of x_0 such that $\mathscr{U}(x) \subset M$ for all $x \in V$, respectively such that $\mathscr{U}(x) \cap M \neq \emptyset$ for all $x \in V$;

(ii) *upper Hausdorff semi-continuous* (u.H.s.c.) *at* x_0, respectively *lower Hausdorff semi-continuous* (l.H.s.c.) *at* x_0, if the relation $\lim_{n\to\infty} x_n = x_0$ implies $\lim_{n\to\infty} \sup_{g \in \mathscr{U}(x_n)} \rho(g, \mathscr{U}(x_0)) = 0$, respectively $\lim_{n\to\infty} \sup_{g \in \mathscr{U}(x_0)} \rho(g, \mathscr{U}(x_n)) = 0$.

The mapping $\mathscr{U}: E \to 2^G$ is called *τ.s.c.* (on E), where $\tau =$ u., u.H., l. or l.H., if it is τ.s.c at each $x_0 \in E$. For u.s.c., respectively, l.s.c., this is clearly equivalent to the usual condition that the set $\{x \in E | \mathscr{U}(x) \subset M\}$, respectively the set $\{x \in E | \mathscr{U}(x) \cap M \neq \emptyset\}$ be open for each open subset M of G. One can show [51, Lemma 1] that $\mathscr{U}$ is u.H.s.c. at x_0 if and only if the relations $\lim_{n\to\infty} x_n = x_0$ and $y_n \in \mathscr{U}(x_n)$, $n = 1, 2, \cdots$, imply $\lim_{n\to\infty} \rho(y_n, \mathscr{U}(x_0)) = 0$ and [174, Lemma 4] that $\mathscr{U}$ is l.s.c. at x_0 if and only if it is l.K.s.c. at x_0. Furthermore (see, for example [148, pp. 20–24]), there exist natural topologies on $2^G \setminus \{\emptyset\}$ such that a mapping $\mathscr{U}: E \to 2^G \setminus \{\emptyset\}$ is *τ.s.c.* at $x_0 \in E$, where $\tau =$ u., u.H., l. or l.H. if and only if it is continuous at x_0 for $2^G \setminus \{\emptyset\}$ endowed with the respective topology. Also, a mapping $\mathscr{U}$ from E into the closed and bounded subsets of G is *Hausdorff continuous at* x_0, i.e., both u.H.s.c. and l.H.s.c. at x_0 if and only if it is continuous at x_0 relative to the Hausdorff metric, that is, the relation $\lim_{n\to\infty} x_n = x_0$ implies

$$\lim_{n\to\infty} \max \left\{ \sup_{g_0 \in \mathscr{U}(x)} \rho(g_0, \mathscr{U}(x_n)), \sup_{g \in \mathscr{U}(x_n)} \rho(g, \mathscr{U}(x_0)) \right\} = 0.$$

Let us also mention that $\mathscr{U}: E \to 2^G$ is called *continuous at* x_0 if it is both u.s.c. and l.s.c. at x_0.

In the particular case when $\mathscr{U}(x)$ is a singleton for each $x \in E$, the above notions reduce to the usual continuity of $\mathscr{U}$. The concepts of upper and lower semi-continuity and Hausdorff continuity go back at least to the 1920's (see Hahn [77] and the bibliography cited there; for a more recent survey, see [52]), while upper and lower Hausdorff semi-continuity were introduced and studied by W. Pollul [148] (see also [136], [51]). Recently B. Brosowski and F. Deutsch [23], [25] have introduced some other notions of "radial" semi-continuity (i.e., along certain lines); however, their results show that these notions are not suitable for the study of metric projections onto linear subspaces (being too weak), but only onto more general subsets of normed linear spaces (see § 5).

It is easy to see [148], [136], [51] that if $\mathscr{U}: E \to 2^G$ is u.s.c. at x_0, then $\mathscr{U}$ is u.H.s.c. at x_0 and that conversely, if $\mathscr{U}$ is u.H.s.c. at x_0 and $\mathscr{U}(x_0)$ is compact, then $\mathscr{U}$ is u.s.c. at x_0. For metric projections onto linear subspaces we have the following stronger result.

THEOREM 4.22. *For a linear subspace G of a normed linear space E and for* $x_0 \in E$, *the following statements are equivalent*:

1°. $\mathscr{P}_G$ *is u.s.c. at* x_0.

2°. $\mathscr{P}_G$ *is u.H.s.c. at* x_0 *and* $\mathscr{P}_G(x_0)$ *is compact.*

The global version of the main part of Theorem 4.22 (i.e., that if $\mathscr{P}_G$ is u.s.c., then $\mathscr{P}_G(x)$ is compact for each $x \in E$) has been proved for proximinal subspaces G in [174, Theorem 1]; in the present more general form Theorem 4.22 was given in [51, Theorem 1]. Note that in general Theorem 4.22 is no longer valid for sets G

which are not linear subspaces [51], since if E is an infinite-dimensional normed linear space and $G = \{x \in E| \, \|x\| \geqq 1\}$, then $\mathscr{P}_G$ is u.s.c. at $x_0 = 0$ but $\mathscr{P}_G(0) = \{x \in E| \, \|x\| = 1\}$ is not compact.

Let us also mention the following extension of the global version of Theorem 4.22, due to G. Godini [73, Theorem 1]: *Let E, G be two normed linear spaces and $\mathscr{U}: E \to 2^G \setminus \{\emptyset\}$ such that $\mathscr{U}(x)$ is bounded for each $x \in E$ and that $\mathscr{U}(\lambda x) = \lambda \mathscr{U}(x)$ for all $\lambda \geqq 1$. Then $\mathscr{U}$ is u.s.c. if and only if* (i) *for each compact set $A \subset E$, the set $\mathscr{U}(A) = \bigcup_{x \in A} \mathscr{U}(x) \subset G$ is compact and* (ii) *the relation $\lim_{n \to \infty} x_n = x_0$ implies $\bigcap_{n=1}^{\infty} \mathscr{U}(x_n) \subset \mathscr{U}(x_0)$.* In particular, if G is a proximinal subspace of E, then $\mathscr{P}_G$ satisfies (ii) (since it is even u.K.s.c., as observed above), hence *$\mathscr{P}_G$ is u.s.c. if and only if for each compact set $A \subset E$, $\mathscr{P}_G(A)$ is compact*; also, G. Godini has observed [73] that if in addition $E = G \oplus F$ (e.g., if codim $G < \infty$), then this is equivalent to the same condition for all compact sets $A \subset F$.

For lower semi-continuity the easier implications are reversed, namely [148], [51], if $\mathscr{U}: E \to 2^G$ is l.H.s.c. at x_0, then $\mathscr{U}$ is l.s.c. at x_0 and conversely, if $\mathscr{U}$ is l.s.c. at x_0 and $\mathscr{U}(x_0)$ is compact, then $\mathscr{U}$ is l.H.s.c. at x_0. However [148, p. 61] *it is not known* any example of a l.s.c. metric projection $\mathscr{P}_G$ which is not l.H.s.c. G. Godini has observed [73] that an u.K.s.c. mapping $\mathscr{U}: E \to 2^G \setminus \{\emptyset\}$ (in particular, $\mathscr{P}_G$ onto a proximinal subspace) is l.s.c. if and only if $\overline{\mathscr{U}(A)} = \mathscr{U}(\bar{A})$ for each conditionally compact set A (i.e., such that $\bar{A}$ is compact); again, if $E = G \oplus F$ and $\mathscr{U} = \mathscr{P}_G$, then it suffices [73] to have this for all conditionally compact sets $A \subset F$.

(d) There arises naturally the problem of the characterization of those linear (and, in particular of those proximinal) subspaces G of a normed linear space E, for which $\mathscr{P}_G$ is τ.s.c., where $\tau = u.$, u.H., l. or l.H. We shall give now several such characterizations (in terms of similar properties of some other related mappings), which extend to set-valued $\mathscr{P}_G$ some of the results of § 4.2. on the continuity of π_G for Chebyshev subspaces G.

For any proximinal subspace G of a normed linear space E one can define [174] a set-valued mapping $\mathscr{V}_G$ of E/G into $2^{\mathscr{P}_G^{-1}(0)}$ by

$$\mathscr{V}_G(x + G) = x - \mathscr{P}_G(x) = \{x - g_0 | g_0 \in \mathscr{P}_G(x)\}, \quad x + G \in E/G. \tag{4.33}$$

Observe that $\mathscr{V}_G$ is well-defined, because of (4.31) and since G is proximinal. It is is easy to see that $\mathscr{V}_G(x + G)$ is a nonvoid closed subset of $\mathscr{P}_G^{-1}(0)$ and that [174]

$$\mathscr{V}_G = (\omega_G|_{\mathscr{P}_G^{-1}(0)})^{-1}. \tag{4.34}$$

THEOREM 4.23. *Let E be a normed linear space, G a proximinal subspace of E, $x_0 \in E$ and τ any of the properties* u., u.H., l. *or* l.H. *The following statements are equivalent*:

1°. *$\mathscr{P}_G$ is τ.s.c. at x_0.*

2°. *$I - \mathscr{P}_G$ is τ.s.c. at x_0.*

3°. *$\mathscr{V}_G$ is τ.s.c at $x_0 + G$.*

Consequently, the corresponding "global" properties (i.e., at all $x_0 \in E$ and $x_0 + G \in E/G$) are also equivalent.

This generalization and localization of Theorem 4.2 has been given in [51, Theorem 2] (for the global equivalence 1° ⇔ 3° for τ = u. or l. see [174, Theorems 3 and 5]). The proof is based on the set-valued extension of the diagram (4.10). For the global properties we also have that $\mathscr{P}_G$ *is u.s.c.* (*respectively l.s.c.*) *if and only if* $\omega_G|_{\mathscr{P}_G^{-1}(0)}$ *carries closed* (*respectively open*) *sets onto closed* (*respectively open*) *sets* [174, Theorems 3 and 5]). Recently G. Godini has observed that this is also equivalent to the condition that for every closed (respectively open) set $A \subset G$, the set $A + \mathscr{P}_G^{-1}(0)$ is closed (respectively open) or to the condition that for every closed (respectively open) set $A \subset \mathscr{P}_G^{-1}(0)$ the set $A + G$ is closed (respectively open).

COROLLARY 4.4. *Let E be a normed linear space and* τ *any one of the properties* u., u.H., l. *or* l.H. *Then*

(i) *For a* $\sigma(E^*, E)$*-closed subspace* Γ *of* E^* *and* $f_0 \in E^*$, $\mathscr{P}_\Gamma$ *is* τ*.s.c. at* f_0 *if and only if the Hahn–Banach extension map*

$$(4.35) \qquad \mathscr{H}: \varphi \in (\Gamma_\perp)^* \to \{f \in E^* | f|_{\Gamma_\perp} = \varphi, \|f\| = \|\varphi\|\} \in 2^{E^*}$$

is τ*.s.c. at* $f_0|_{\Gamma_\perp}$. *Consequently, the corresponding global properties are also equivalent.*

(ii) *If for a proximinal subspace G of E and* $x_0 \in E$ *the mapping* $\delta \circ \mathscr{H}_1 : (G^\perp)^* \to 2^E$ *is* τ*.s.c. at* $\kappa(x_0)|_{G^\perp}$, *where* $\mathscr{H}_1$ *is the Hahn–Banach extension map*

$$(4.36) \qquad \mathscr{H}_1 : f \in (G^\perp)^* \to \{\Phi \in E^{**} | \Phi|_{G^\perp} = f, \|\Phi\| = \|f\|\} \in 2^{E^{**}}$$

and where $\delta : 2^{E^{**}} \to 2^E$ *is defined by*

$$(4.37) \qquad \delta(A) = \kappa^{-1}(\kappa(E) \cap A), \qquad A \in 2^{E^{**}},$$

then $\mathscr{P}_G$ *is* τ*.s.c. at* x_0 *and, if* E/G *is reflexive, then the converse is also true. Consequently, for the corresponding global statements the same implications hold.*

This generalization of Corollary 4.1 has been given in [51, Theorems 5 and 6] (for the global part of (i) for $\tau =$ u. or l. see [174, Corollaries 4 and 5]). The proofs are based on the set-valued extensions of the diagrams (4.11) (or (4.12)) and (4.14).

(e) No theorem is known in concrete spaces about characterization of subspaces G of arbitrary dimension with a τ.s.c. metric projection $\mathscr{P}_G$, where $\tau =$ u., u.H., l. or l.H.

(f) Let us consider now, in arbitrary normed linear spaces, the problem of characterization of linear subspaces G with τ.s.c. $\mathscr{P}_G$, when there are restrictions on $\dim G$ or $\operatorname{codim} G$, or restrictions on $\dim \mathscr{P}_G(x)$, $x \in E$. We have the following extension of Theorem 4.3 for $\tau =$ u., u.H.

THEOREM 4.24. *For every finite-dimensional linear subspace G of a normed linear space E,* $\mathscr{P}_G$ *is u.s.c.* (*hence also u.H.s.c.*).

Indeed, this is a particular case of the following more general result (see [168, p. 386]): *If G is an "approximatively compact" set in a metric space E* (*i.e., for every* $x \in E$ *and* $\{g_n\} \subset G$ *with* $\lim_{n\to\infty} \rho(x, g_n) = \rho(x, G)$ *there exists a subsequence* $\{g_{n_k}\}$ *converging to an element of G*), *then* $\mathscr{P}_G$ *is u.s.c.*

In other words, Theorem 4.24 says that *for any superspace E of a finite-dimensional normed linear space G,* $\mathscr{P}_G$ *is u.s.c.* In [51, Theorem 3] it was shown that this property actually characterizes finite-dimensional normed linear spaces G, namely *every infinite-dimensional normed linear space G can be embedded isometrically, as a proximinal hyperplane* G_0, *into a normed linear space E, in such a way that*

$\mathscr{P}_{G_0}$ *is not u.s.c.* Indeed, one can take $E = G \times K$ (where K is the field of scalars) endowed with the norm $\|\{g, \lambda\}\| = \max(\|g\|, |\lambda|)$, and $G_0 = G \times \{0\}$ (by Theorem 4.22 above). Such a characterization of finite-dimensionality in terms of lower semi-continuity of $\mathscr{P}_G$ is no longer possible, since, on the one hand, there exist even one-dimensional subspaces G of certain spaces E such that $\mathscr{P}_G$ is not l.s.c. (see Theorem 4.31 below) and, on the other hand, for any closed hyperplane G, $\mathscr{P}_G$ is l.s.c. (see Theorem 4.26 below). However [51, Theorem 4], *every real normed linear space G can be embedded isometrically, as a proximinal subspace G_0 of co-dimension 2, into a real normed linear space E, in such a way that $\mathscr{P}_{G_0}$ is not l.s.c.*

For $\tau =$ u. we have the following extension of Theorems 4.5 and 4.4.

THEOREM 4.25. *For a closed linear subspace G of finite codimension of a normed linear space E, the following statements are equivalent*:

1°. *$\mathscr{P}_G$ is u.s.c.*

2°. *$\mathscr{P}_G^{-1}(0)$ is boundedly compact.*

3°. *$\mathscr{P}_G^{-1}(0) \cap \operatorname{Fr} S_E$ is compact.*

These statements imply—and if codim $G = 1$, *they are equivalent to—the following statement*:

4°. *$\mathscr{P}_G(x)$ is compact for every $x \in E$.*

This theorem has been given in [174, Corollary 1 and Theorems 2,3]; a weaker version of the equivalence $1° \Leftrightarrow 2°$ (namely, that $1° \cap 4° \Leftrightarrow 2°$ for codim $G < \infty$) has been given by P. D. Morris [131, Theorem 3]. From the implication $4° \Rightarrow 1°$ it follows [174, Corollary 2] that *if G is a pseudo-Chebyshev hyperplane, then $\mathscr{P}_G$ is u.s.c.* For $\tau =$ u. H., l.H. and l. we have the following extension of the final part of Theorem 4.4, due to W. Pollul [148, p. 61, Theorem 1].

THEOREM 4.26. *For every closed hyperplane G in a normed linear space E, $\mathscr{P}_G$ is uniformly continuous relative to the Hausdorff metric* (whence also u.H.s.c., l.H.s.c. and l.s.c.), *with Lipschitz constant* 2.

For $\tau =$ u.H. and $\tau =$ l. this result has been also obtained by E. V. Oshman [139, Theorem 8] and in [174, Theorem 4] respectively.

We recall that the set-valued "spherical image map" $\mathscr{T}$ from E^* into 2^E is defined by

$$\mathscr{T}(f) = \{x \in E \mid f(x) = \|f\| \|x\|, \|x\| = \|f\|\}, \qquad f \in E^*. \tag{4.38}$$

We have the following extension of Theorem 4.6 for $\tau =$ u.

THEOREM 4.27. *For a subspace G of finite codimension of a normed linear space E the following statements are equivalent*:

1°. *$\mathscr{P}_G$ is u.s.c.*

2°. *$\mathscr{T}|_{\operatorname{Fr} S_{G^\perp}}$ is u.H.s.c. and for each $f \in \operatorname{Fr} S_{G^\perp}$ the set $\mathscr{T}(f)$ is compact.*

3°. *$\mathscr{T}|_{\operatorname{Fr} S_{G^\perp}}$ is u.s.c. and for each $f \in \operatorname{Fr} S_{G^\perp}$ the set $\mathscr{T}(f)$ is compact.*

4°. *$\mathscr{T}|_{G^\perp}$ is u.H.s.c. and for each $f \in G^\perp$ the set $\mathscr{T}(f)$ is compact.*

5°. *$\mathscr{T}|_{G^\perp}$ is u.s.c.*

Theorem 4.27 was given in [51, Theorem 7], where it was also observed that the condition of compactness in 2° – 4° cannot be omitted, as shown by the following example: If G is a proximinal hyperplane in a real normed linear space E such that for some (and hence for each) $x \in E \setminus G$ the set $\mathscr{P}_G(x)$ is not compact, then $\mathscr{P}_G$ is not u.s.c. (by Theorem 4.22) and $\mathcal{T}(f)$ is not compact for any $f \in \operatorname{Fr} S_{G^\perp}$ but $\mathcal{T}|_{\operatorname{Fr} S_{G^\perp}}$ is u.s.c. (hence u.H.s.c.) and $\mathcal{T}|_{G^\perp}$ is u.H.s.c. (but not u.s.c.).

The analogous extension of Theorem 4.6 for $\tau = 1$. is not valid, as shown by the following example [51, Proposition 1]: *If* $3 \leqq \dim E < \infty$ *and if* S_E *is a polyhedron, then for any linear subspace G of E with* $\dim G \leqq \dim E - 2$, $\mathcal{P}_G$ *is l.s.c. but* $\mathcal{T}|_{\operatorname{Fr} S_{G^\perp}}$ *is not l.s.c.* However, the implication in the other direction is always valid:

THEOREM 4.28. *Let G be a subspace of finite codimension of a normed linear space E. If* $\mathcal{T}|_{\operatorname{Fr} S_{G^\perp}}$ *is l.s.c., then* $\mathcal{P}_G$ *is l.s.c.*

Theorem 4.28 was given in [51, Theorem 9]; in the particular case when $\dim E < \infty$, it was proved by R. Wegmann [190].

Finally, let us also mention the following sufficient condition for the continuity of u.s.c. metric projections onto pseudo-Chebyshev subspaces, due to B. Brosowski and R. Wegmann [29, Theorem 3].

PROPOSITION 4.9. *Let G be a pseudo-Chebyshev subspace of a normed linear space E, such that* $\mathcal{P}_G$ *is u.s.c. If for each* $x \in \mathcal{P}_G^{-1}(0)$ *there exists a* $\sigma(E^*, E)$*-open set* $M \subset E^*$ *such that*

$$\bigcap_{g_0 \in \mathcal{P}_G(x)} \{f \in \mathcal{E}(S_{E^*}) | f(x - g_0) = \|x - g_0\|\} \subset M \cap \mathcal{E}(S_{E^*})$$

$$\subset \bigcap_{g_0 \in \mathcal{P}_G(x)} \{f \in E^* | f(g_0) = 0\},$$

then $\mathcal{P}_G$ *is l.s.c. (and hence continuous).*

(g) For the spaces $E = C_R(Q)$ we have the following results, the first one due to P. D. Morris [131], the second to J. Blatter, P. D. Morris and D. E. Wulbert [15] (recently, their proof has been simplified by A. L. Brown [33]) and the third to B. Brosowski and R. Wegmann [29] (recently, their proof has been simplified by W. Nitzsche [133]).

THEOREM 4.29. (i) *If G is a pseudo-Chebyshev subspace of finite codimension* $n \geqq 2$ *of* $E = C_R(Q)$ *(Q compact), then* $\mathcal{P}_G$ *is not u.s.c.*

(ii) *For a pseudo-Chebyshev subspace G of* $E = C_R(Q)$, *in order that* $\mathcal{P}_G$ *be l.s.c. it is necessary, and if* $\mathcal{P}_G$ *is u.s.c., also sufficient, that for every* $x \in \mathcal{P}_G^{-1}(0)$ *the set* $\bigcap_{g_0 \in \mathcal{P}_G(x)} Z(g_0)$ *be open* (where $Z(g_0) = \{q \in Q | g_0(q) = 0\}$.

(iii) *Part* (ii) *remains valid for* $E = (C_0)_F(T)$, *where T is a locally compact space and F a strictly convex normed linear space.*

Part (i) is a generalization of Theorem 4.7 for $\tau = u$. Blatter, Morris and Wulbert [15] have proved the following corollary of part (ii).

COROLLARY 4.5. *For a compact space Q the following conditions are equivalent:*

1°. *Q is connected.*

2°. *Every pseudo-Chebyshev subspace G of* $E = C_R(Q)$, *such that* $\mathcal{P}_G$ *is l.s.c., is a Chebyshev subspace.*

3°. *Every one-dimensional subspace G of* $E = C_R(Q)$, *such that* $\mathcal{P}_G$ *is l.s.c., is a Chebyshev subspace.*

For some related results see also B. Brosowski, K.-H. Hoffmann, E. Schäfer and H. Weber [27].

For the spaces $E = L_R^1(T, \nu)$, A. Lazar, D. E. Wulbert and P. D. Morris [121] have proved (and recently B. Brosowski and R. Wegmann [29] have given a different proof of the sufficiency part, using Proposition 4.10):

THEOREM 4.30. *For an n-dimensional linear subspace G of* $E = L_R^1(T, \nu)$, *where* (T, ν) *is a* σ*-finite positive measure space,* $\mathscr{P}_G$ *is l.s.c. if and only if there do not exist* $\beta \in G^{\perp}\setminus\{0\}$ *and* $g \in G$ *with the following three properties*:

(i) *The set* $S(\beta) = \{t \in T \mid |\beta(t)| < \|\beta\|\}$ *is purely atomic and contains at most* $n - 1$ *atoms.*

(ii) $S(\beta) \subset Z(g)$.

(iii) $T\setminus Z(g)$ *is not the union of a finite family of atoms.*

From this result one obtains [121]

COROLLARY 4.6. *If G is a finite-dimensional non-Chebyshev linear subspace of* $E = L_R^1(T, \nu)$ (*where* (T, ν) *is a* σ*-finite positive measure space*) *such that no* $g \in G\setminus\{0\}$ *has* $T\setminus Z(g)$ *purely atomic, consisting of a finite number of atoms* (*in particular, if* (T, ν) *has no atoms and hence any finite dimensional G is non-Chebyshev by* § 3, *Theorem* 3.12), *then* $\mathscr{P}_G$ *is not l.s.c.*

For subspaces which can be of infinite dimension and infinite codimension we have the following result of J. Blatter [13, Theorem 3.12]: *If G is a Grothendieck subspace of* $E = C_F(Q)$, *where Q is compact and* $F^* \equiv L^1(T, \nu)$ *for some positive measure space* (T, ν) *and if G satisfies* § 2, *condition* (2.19), *then* $\mathscr{P}_G$ *is Hausdorff continuous* (whence also u.H.s.c., l.H.s.c. and l.s.c.), *with Lipschitz constant* $2/\lambda$. Hence, in particular, if *G is a Weierstrass–Stone subspace of* $C_F(Q)$, *then* $\mathscr{P}_G$ *has Lipschitz constant* 2 (since we have seen in § 2 that G satisfies (2.19) with $\lambda = 1$). Independently, B. R. Kripke has proved (see [82, p. 173]) that *for the* (*proximinal, non-Chebyshev*) *subspace* $G = l_R^\infty(\Omega) \cap C_R(\Omega)$ *of* $E = l_R^\infty(\Omega)$, *where* Ω *is a paracompact space,* $\mathscr{P}_G$ *is Hausdorff continuous, with Lipschitz constant* 2; however, it turns out (see § 2, the comments after Corollary 2.2) that this is a particular case of the above result of Blatter.

Let us consider now the normed linear spaces in which for every proximinal subspace G the set-valued metric projection $\mathscr{P}_G$ has one of the above semi-continuity or continuity properties. E. V. Oshman has proved [139, Theorem 1] that *every closed linear subspace G of a Banach space E is proximinal and with* $\mathscr{P}_G$ *u.s.c. if and only if G is reflexive and the relations* $\{x_n\}_0^\infty \subset E$, $\{f_n\}_0^\infty \subset E^*$, $f_n(x_n) = \|x_n\| = \|f_n\| = 1$, $n = 0, 1, 2, \cdots$, $\lim_{n\to\infty} \rho(x_n - x_0, [f_0, f_1, f_2, \cdots]_\perp) = 0$ *and* $x_n \to x_0$ *weakly, imply* $\lim_{n\to\infty} x_n = x_0$. More generally, E. V. Oshman has observed [139, Theorem 6] that *if a* (not necessarily reflexive) *normed linear space E satisfies the last condition, then for every reflexive subspace G of E,* $\mathscr{P}_G$ *is u.H.s.c.* Actually, one can also show that in this case $\mathscr{P}_G$ *is u.s.c.* and that, *conversely, if for every reflexive subspace G of E,* $\mathscr{P}_G$ *is u.s.c., then E satisfies the above condition with* $[f_0, f_1, f_2, \cdots]_\perp$ *reflexive.* Indeed, this results from the following fact: *In order that for a proximinal subspace G of a normed linear space, E,* $\mathscr{P}_G$ *be u.s.c., it is necessary, and if G is reflexive, also sufficient, that the relations* $x_n \in \mathscr{P}_G^{-1}(0)$, $n = 1, 2, \cdots$, $\lim_{n\to\infty} \rho(x_n - x_0, G) = 0$ *and* $x_n \to x_0$ *weakly, imply* $\lim_{n\to\infty} x_n = x_0$ (compare this with the reformulation of Theorem 4.2 given before Corollary 4.1; note also that now $x_0 \in \mathscr{P}_G^{-1}(0)$ is not assumed, but it follows from the other assumptions). In [166] (see also [175, Theorem 4.25]) it was observed (and it can be also deduced from the above theorem of Oshman) that we have the following extension of a remark made after Proposition 4.2 above: *For all closed linear subspaces G of a reflexive Banach space E with property* (H), $\mathscr{P}_G$ *is u.s.c.*

There also exist such results when we have restrictions on dim G or codim G. Thus, A. L. Brown [32] has proved the equivalence 1° ⇔ 3° (and introduced "property (P)") and Blatter, Morris and Wulbert [15] have observed the other equivalences of the next theorem.

THEOREM 4.31. *For a normed linear space E the following statements are equivalent*:

1°. *For every finite-dimensional linear subspace G of E,* $\mathscr{P}_G$ *is l.s.c.*

2°. *For every one-dimensional linear subspace G of E,* $\mathscr{P}_G$ *is l.s.c.*

3°. *E has the following property*:

(P) *For every pair of elements* $x, z \in E$ *such that* $\|x + z\| \leqq \|x\|$, *there exist constants* $b = b(x, z) > 0$, $c = c(x, z) > 0$ *such that*

$$\|y + cz\| \leqq \|y\|, \qquad y \in S(x, b). \tag{4.39}$$

It is natural to ask, which normed linear spaces E have property (P). A. L. Brown [32] has proved (i) and (ii), and Blatter [9] and Blatter–Morris–Wulbert [15] have proved the other statements of

PROPOSITION 4.10. (i) *Every strictly convex normed linear space E has property* (P).

(ii) *Every finite-dimensional normed linear space E, in which the unit cell* S_E *is a polyhedron, has property* (P).

(iii) $C_R(Q)$ *(Q compact) has property* (P) *if and only if Q is finite.*

(iv) c_0 *has property* (P).

(v) *If* (T, ν) *is a* σ*-finite measure space such that T is not the union of a finite number of atoms, then* $L^1(T, \nu)$ *does not have property* (P).

E. V. Oshman has proved [139, Theorem 2] that *in a reflexive Banach space E the metric projection* $\mathscr{P}_G$ *onto every closed linear subspace G of finite codimension (respectively, of codimension* $\leqq k < \infty$*) is u.s.c. if and only if the relations* $\{x_n\}_0^\infty \subset E$, $\{f_n\}_0^\infty \subset E^*$, $f_n(x_n) = \|x_n\| = \|f_n\| = 1, n = 0, 1, 2, \cdots$, $x_n \to x_0$ *weakly and* $\dim [f_n]_0^\infty < \infty$ *(respectively,* $\dim [f_n]_0^\infty \leqq k$*) imply* $\lim_{n\to\infty} x_n = x_0$. This extends the result of Oshman given after Proposition 4.3 above.

Finally, we mention that for a normed linear space E the following statements are equivalent: 1°. *For every closed hyperplane G in E,* $\mathscr{P}_G$ *is u.s.c.* 2°. *Every maximal convex subset of* Fr S_E *is compact.* 3°. *For every* $f \in \text{Fr } S_{E^*}$ *the set* $\{x \in E \mid f(x) = 1\} \cap \text{Fr } S_E$ *is either void or compact.* Indeed, the equivalence 1° ⇔ 2° was proved in [174, Corollary 3] and the equivalence 1° ⇔ 3° has been given, independently, by E. V. Oshman [139, Theorem 9].

It is also natural to raise [174] the problem of extending the results of §§ 4.3–4.6 to set-valued metric projections, for example, to study weakly semi-continuous $\mathscr{P}_G$ and those $\mathscr{P}_G$ which are Lipschitzian for the Hausdorff metric on $2^G \setminus \{\varnothing\}$, or to find a suitable generalization of linearity for set-valued $\mathscr{P}_G$ and then give characterizations of those linear subspaces G for which $\mathscr{P}_G$ has this property (e.g., in terms of a suitable property of $\mathscr{V}_G$).

4.8. Continuous selections and linear selections for set-valued metric projections. We recall that if G and E are metric spaces, a continuous mapping $u: E \to G$ is said to be a *continuous selection* for a set-valued mapping $\mathscr{U}: E \to 2^G \setminus \{\emptyset\}$ if $u(x) \in \mathscr{U}(x)$ for all $x \in E$. If G is a linear subspace of a normed linear space E, one can define a *linear selection* for $\mathscr{U}$ in a similar way. By Theorem 4.1 (ii) or (iii), if G is proximinal, every linear selection for $\mathscr{P}_G$ is continuous.

We have the following generalization of Theorem 4.2 (which, in the particular case when E/G is reflexive, was essentially proved in [170, Theorem 3] and in the general case in [174, Theorem 6]).

THEOREM 4.32. *For a proximinal linear subspace G of a normed linear space E, $\mathscr{P}_G: E \to 2^G$ admits a continuous selection if and only if the mapping $\mathscr{V}_G: E/G \to 2^{\mathscr{P}_G^{-1}(0)}$ defined by* (4.33) *admits a continuous selection.*

Indeed, this can be proved either similarly to the above proof of Theorem 4.2, or using, in the necessity part, a theorem of Bartle and Graves (see [128]) according to which the mapping $\mathscr{W}_G: E/G \to 2^E$ defined by

$$\mathscr{W}_G(x + G) = \{x + g \mid g \in G\} = x + G \tag{4.40}$$

always admits a continuous selection w and then putting

$$v_G(x + G) = w_G(x + G) - \pi_G^{(0)} w_G(x + G), \qquad x + G \in E/G, \tag{4.41}$$

where $\pi_G^{(0)}$ is a continuous selection for $\mathscr{P}_G$.

From Theorem 4.32 we obtain the following generalization of Corollary 4.1, the first part of which is due to J. Lindenstrauss [122] and the second part to [170].

COROLLARY 4.7. *Let E be a normed linear space. Then*

(i) *For a $\sigma(E^*, E)$-closed linear subspace Γ of E^*, $\mathscr{P}_\Gamma$ admits a continuous selection $\pi_\Gamma^{(0)}$ if and only if the Hahn–Banach extension map* (4.35) *admits a continuous selection.*

(ii) *If G is a proximinal linear subspace of E, such that the Hahn–Banach extension map* (4.36) *admits a continuous selection, then $\mathscr{P}_G$ admits a continuous selection.*

The results of § 4.7 on lower semi-continuity are particularly useful because of the following theorem on continuous selections, due to E. A. Michael [128, Theorem 3.2″]. *If E, G are Banach spaces, every l.s.c. mapping $\mathscr{U}: E \to 2^G \setminus \{\emptyset\}$ such that $\mathscr{U}(x)$ is closed and convex for each $x \in E$, admits a continuous selection.* Hence, in particular, *if G is a proximinal linear subspace of a Banach space E, such that $\mathscr{P}_G$ is l.s.c.* (or, in particular, Hausdorff continuous), *then $\mathscr{P}_G$ admits a continuous selection*; the converse is not true, even if $\dim G = 1$. From this observation and from the results of § 4.7 there follow sufficient conditions on a given G in order that $\mathscr{P}_G$ admit a continuous selection and sufficient conditions on E in order that for all finite-dimensional subspaces G of E, $\mathscr{P}_G$ admit a continuous selection. Conversely, in some cases the results on non-lower semi-continuity of $\mathscr{P}_G$ can be sharpened to nonexistence of continuous selections for $\mathscr{P}_G$. For example, A. Lazar, D. E. Wulbert and P. D. Morris have proved the following partial sharpening of Corollary 4.6 [121, Theorem 1.4].

THEOREM 4.33. *If G is any finite-dimensional subspace of $E = L_R^1(T, \nu)$, where (T, ν) is a positive measure space having no atoms, then $\mathscr{P}_G$ admits no continuous selection.*

We have the following characterizations of the one-dimensional linear subspaces G of $E = l_R^1$ and $E = C_R(Q)$ for which $\mathscr{P}_G$ admits a continuous selection, due to A. Lazar [120, Lemma 5.2] and respectively A. Lazar, D. E. Wulbert and P. D. Morris [121, Proposition 2.6].

PROPOSITION 4.11. (i) *For the one-dimensional linear subspace $G = [g]$ of $E = l_R^1$ spanned by an element $g = \{\gamma_n\}$, $\mathscr{P}_G$ admits a continuous selection if and only if there do not exist two disjoint subsets N_1, N_2 of $\mathscr{N} = \{1, 2, 3, \cdots\}$ such that*

$$(\alpha) \qquad N_1 \cup N_2 = \{i \in \mathscr{N} | \gamma_i \neq 0\},$$

$$(\beta) \qquad \sum_{i \in N_1} |\gamma_i| = \sum_{i \in N_2} |\gamma_i|.$$

(ii) *For the one-dimensional linear subspace $G = [g]$ of $E = C_R(Q)$ (Q compact), spanned by an element g, $\mathscr{P}_G$ admits a continuous selection if and only if*

(α) Fr $Z(g)$ *is either void or a singleton.*

(β) $q \in$ Fr $Z(g)$ *implies there exists a neighborhood of q on which g is either nonpositive or nonnegative.*

Let us also mention the following results of A. L. Brown [33, Theorems 2.8 and 3.10].

THEOREM 4.34. (i) *If G is a pseudo-Chebyshev "Z-subspace" of $E = C_R(Q)$ (Q compact), i.e., a pseudo-Chebyshev subspace such that* Int $Z(g) = \varnothing$ *for all $g \in G \setminus \{0\}$, then either there is no continuous selection for $\mathscr{P}_G$ or there is a unique one.*

(ii) *There exists a 5-dimensional Z-subspace G of $E = C_R([-1, +1])$ which contains the constants, is non-Chebyshev and such that $\mathscr{P}_G$ admits a unique continuous selection.*

The latter result (which disproves a claim of A. Lazar, D. E. Wulbert and P. D. Morris: [121, Theorem 2.1]) shows that in the particular case when $\dim G < \infty$ (and hence $\mathscr{P}_G$ is u.s.c.), the implication $1° \Rightarrow 2°$ of Corollary 4.5 cannot be sharpened so as to assume only existence of a continuous selection for $\mathscr{P}_G$ instead of the lower semi-continuity of $\mathscr{P}_G$.

Concerning linear selections for $\mathscr{P}_G$ we have (see [168, p. 142]):

THEOREM 4.35. *For every proximinal hyperplane G in a normed linear space E, $\mathscr{P}_G$ admits a linear selection.*

On the other hand, the argument at the end of § 4.6(b) shows that *if G is a finite-dimensional subspace of $E = C_R(Q)$, where Q is a perfect compact space and F a Banach space, then $\mathscr{P}_G$ admits no linear selection.*

By Theorem 4.1, if for a proximinal linear subspace G of a normed linear space E, $\mathscr{P}_G$ admits a linear selection $\pi_G^{(0)}$, then $\pi_G^{(0)}$ is a continuous linear projection of E onto G and hence G *is complemented* (obviously, the converse is not valid);

moreover, in this case $\|I - \pi_G^{(0)}\| = 1$. Conversely, *if p is a linear projection of E onto G such that $\|I - p\| = 1$, then $p = \pi_G^{(0)}$*, a linear selection for $\mathscr{P}_G$, since $\|x - p(x)\| = \|x - g - p(x - g)\| \leqq \|x - g\|$ for all $g \in G$ [44, Lemma 2]. Also, if *p is a linear projection of $E = C_F(Q)$ onto a subspace G of finite codimension* (where *Q,F are as above*), *we have $p = \pi_G^{(0)}$ (a linear selection for $\mathscr{P}_G$) if and only if $\|p\| = 2$* [44]; indeed, since $I - p$ is compact, we have [64] $\|p\| = \|I - (I - p)\| = 1 + \|I - p\|$, whence the assertion follows.

Finally, let us mention that one can give a characterization of proximinal linear subspaces G for which $\mathscr{P}_G$ admits a linear selection, generalizing Theorem 4.18 in a similar way as we generalized Theorem 4.2 by Theorem 4.32 and one can then prove also a corollary corresponding to Corollary 4.7. Also, one can define weakly continuous, Lipschitzian and differentiable selections for $\mathscr{P}_G$ and obtain for them similar extensions of the preceding results.

5. Best approximation by elements of nonlinear sets.

5.1. Best approximation by elements of convex sets. Extensions to convex optimization in locally convex spaces. By a *nonlinear set* in a normed linear space E we mean any set $G \subset E$ which is not a "linear manifold", i.e., which is not of the form $x + G_0$, where $x \in E$ and where G_0 is a linear subspace of E. Since best approximation by elements of linear manifolds can be reduced, by a simple translation, to best approximation by elements of linear subspaces, we shall not consider here this problem, but refer the reader to [168, pp. 135–140 and 242–246]. We want to present here, briefly, some results and directions of research on best approximation by elements of nonlinear sets. Note that the existing results in this field do not yet constitute a unified theory (as is the theory of best approximation by elements of linear subspaces) and the construction of such a theory in general normed linear spaces is only at its beginnings.

(a) The first natural step when passing from best approximation in normed linear spaces E by elements of linear sets $G \subset E$ to nonlinear sets is to take as G a convex set in E. The following extension of § 1, Theorem 1.1, to this case has been given, for real scalars by G. Sh. Rubinstein [156] and Ch. Roumieu [155, Proposition 5] and for complex scalars in [168, pp. 360–361] and [49], [79] (independently).

THEOREM 5.1. *Let G be a convex set in a normed linear space E, and let $x \in E \setminus \bar{G}$, $g_0 \in G$. We have $g_0 \in \mathscr{P}_G(x)$ if and only if there exists an $f \in E^*$ with the following properties*:

$$\|f\| = 1, \tag{5.1}$$

$$\operatorname{Re} f(g_0 - g) \geqq 0, \qquad g \in G, \tag{5.2}$$

$$\operatorname{Re} f(x - g_0) = \|x - g_0\|. \tag{5.3}$$

This theorem admits the following geometric interpretation, observed by V. N. Burov (see [168, p. 362]): $g_0 \in \mathscr{P}_G(x)$ *if and only if there exists a real hyperplane H which separates G from $S(x, \|x - g_0\|)$*. Clearly, such a hyperplane H must pass through g_0 and support the cell $S(x, \|x - g_0\|)$. The particular case of Theorem 5.1, when G is a convex cone, was also considered by G. Sh. Rubinstein (see [168, pp. 362–363]); another characterization theorem for best approximation by elements of convex cones has been given by G. Godini [72].

For finite-dimensional convex sets F. R. Deutsch and P. H. Maserick [49] and, independently, S. Ia. Havinson [79] have proved the following extension of § 1, Theorem 1.10.

THEOREM 5.2. *Let G be an n-dimensional convex set in a normed linear space E and let $x \in E \setminus G$, $g_0 \in G$. We have $g_0 \in \mathcal{P}_G(x)$ if and only if there exist h extremal points $f_1, \cdots, f_h$ of S_{E^*}, where $1 \leqq h \leqq n + 1$ if the scalars are real and $1 \leqq h \leqq 2n + 1$ if the scalars are complex, and h numbers $\lambda_1, \cdots, \lambda_h > 0$ with $\sum_{j=1}^h \lambda_j = 1$, such that*

$$\text{Re} \sum_{j=1}^{h} \lambda_j f_j(g_0 - g) \geqq 0, \qquad g \in G, \tag{5.4}$$

$$\text{Re} \sum_{j=1}^{h} \lambda_j f_j(x - g_0) = \|x - g_0\|. \tag{5.5}$$

Actually, this follows from Theorem 5.1 in the same way as § 1, Theorem 1.10 follows from § 1, Theorem 1.1.

The second main characterization theorem of § 1 (Theorem 1.2) remains valid in the case when G is a convex set in E; this was observed by A. L. Garkavi (see [168, p. 360]) and, independently, by G. Choquet (unpublished).

Some other characterizations of elements of best approximation by elements of convex sets G have been obtained by P. J. Laurent [116] (see also [119]), who has used the "cones of displacement" (introduced by A. I. Dubovitzkii and A. A. Miliutin [54]) and by J. J. Moreau [130], who has used the tools of the theory of convex functionals (e.g., subdifferentials, infimal convolutions, etc.; see [129] and [95]).

(b) It was observed in [168, p. 360], that several results on existence and uniqueness of best approximation (for example, § 2, Theorem 2.8, with S_G replaced by "all bounded subsets of G"; § 2, Theorem 2.14 with the following addition: 6°. All closed convex subsets of E are proximinal; § 3, Theorem 3.11 with the following addition: 4°. All closed convex subsets of E are Chebyshev sets) and on properties of the mappings π_G not involving the linearity of G, remain valid for the case when G is a convex set in E. However, a systematic extension of the results on best approximation by elements of linear subspaces G to convex sets G has not yet been accomplished. We mention that some results on existence and uniqueness of elements of best approximation have been extended to *convex cones* by G. Godini [72]. Some difficulties which arise at the best approximation by elements of certain finite-dimensional convex cones in $E = C_R([0, 1])$ have been pointed out by J. R. Rice (see [168, p. 363]); for example, if $x_1, \cdots, x_n$ is a Chebyshev system in $E = C_R([0, 1])$, with $n \geqq 2$ and if

$$G = \left\{ \sum_{i=1}^{n} \alpha_i x_i \,\middle|\, \sum_{i=1}^{n} \alpha_i x_i(t) \geqq 0 \; (t \in [0, 1]) \right\}, \tag{5.6}$$

then the characterization of the elements of best approximation is complicated and the problem of their uniqueness is still unsolved. P. R. Halmos [78] has proved that *the convex cone* $G = \Pi(H, H)$ *of all positive continuous linear operators* v *on a Hilbert space* H (i.e., such that $(v(y), y) \geqq 0$ for all $y \in H$) *is proximinal in* $E = L(H, H)$ and has computed the distance $\rho(u, G)$, $u \in E$.

(c) There also exist some results on metric projections onto closed convex sets. E. V. Oshman has proved [135], [140] that *every closed convex set* G *in a Banach space* E *is a Chebyshev set with continuous metric projection* π_G *if and only if* E *is reflexive and the relations* $\{x_n\}_0^\infty \subset E$, $\{f_n\}_0^\infty \subset E^*$, $f_n(x_n) = \|x_n\| = \|f_n\| = 1$, $n = 0, 1, 2, \cdots$, $f_n \to f_0$ *for* $\sigma(E^*, E)$, *and* $\lim_{n\to\infty} \rho(x_n, G_{f_0} \cap G_{f_n}) = 0$, $\lim_{n\to\infty} \rho(x_0, G_{f_0} \cap G_{f_n}) = 0$, *where* $G_{f_n} = \{x \in E | f_n(x) = 1\}$, $n = 0, 1, 2, \cdots$, *imply* $\lim_{n\to\infty} x_n = x_0$. Also, Oshman has observed [139] that the reflexive space l_R^2 *admits a strictly convex isomorph* E *such that for every closed linear subspace* $G \subset E$ *of finite codimension* π_G *is continuous, but for some convex set* $G_0 \subset E$, π_{G_0} *is discontinuous*; namely, it is sufficient to take E to be l_R^2 endowed with the equivalent norm $\|\cdot\|_1$ defined as the Minkowski functional of the set

$$S_1 = \left\{x = \{\xi_n\} \in l_R^2 \,\middle|\, \sum_{i=1}^{\infty} \frac{\xi_i^2}{i^2} \leqq 1, \sum_{i=1}^{\infty} \xi_i^2 \leqq 3\right\}. \tag{5.7}$$

It is well known (see, e.g., [82, pp. 157–158]) that *if* E *is an inner product space and* G *a complete convex set in* E, *then* G *is a Chebyshev set and*

$$\|\pi_G(x) - \pi_G(y)\| \leqq \|x - y\|, \qquad x, y \in E, \tag{5.8}$$

$$\operatorname{Re}(x - y, \pi_G(x) - \pi_G(y)) \geqq \|\pi_G(x) - \pi_G(y)\|^2 \geqq 0, \qquad x, y \in E, \tag{5.9}$$

so π_G is a "contraction" (i.e., Lipschitzian with constant 1) and "monotone" mapping on E, and that either of these properties of π_G characterizes inner product spaces among general normed linear spaces.

It is also well known (see, e.g., [83]) that if G is a closed convex set in a finite-dimensional Euclidean space E then, since π_G is Lipschitzian, π_G is almost everywhere Fréchet differentiable. However, J. Kruskal [112] has constructed a convex subset G of $E = l_3^2 (= R^3)$ such that π_G fails to have a one-sided directional derivative at some points of E.

When $\dim E = \infty$, Lipschitzian mappings need not be differentiable. R. B. Holmes has studied [83] the differential properties of π_G for a closed convex set G in a Hilbert space $E = H$, under appropriate assumptions on the boundary of G; for example [83], if for some $x_0 \in H$ the boundary of G is of class C^{p+1} near $\pi_G(x_0)$, then π_G is of class C^p on a neighborhood of x_0. Also, various properties of the differential operator $D\pi_G(x)$ are obtained in [83]; for example, those cases are characterized when this operator is a constant multiple of the orthogonal projection of H onto the tangent space to G at x. We also mention (see [82], [83], [197]) that *if* G *is a closed convex set in a Hilbert space* $E = H$ *and*

$$\varphi(x) = \tfrac{1}{2}\rho(x, G)^2, \qquad x \in H, \tag{5.10}$$

then φ *is a smooth convex function on* H *and the gradient of* φ *is given by the formula*

$$\nabla\varphi = I - \pi_G. \tag{5.11}$$

For some more results on the differentiability of π_G onto convex sets G in a Hilbert space $E = H$, and some other properties of π_G, see also the work of E. Zarantonello [197].

(d) For set-valued metric projections onto convex sets, R. Wegmann [191] has given an extension of § 4, Theorem 4.31, implication $3° \Rightarrow 1°$. Namely, a convex set G in a normed linear space E is said [191] to have *property* (P), if for every $x \in G$, $z \in E$ with $x + z \in G$ there exist constants $b = b(x, z) > 0$, $c = c(x, z) > 0$ such that $y + cz \in G$ for all $y \in G \cap \operatorname{Int} S(x, b)$; then, E has property (P) of § 4, Theorem 4.31 if and only if S_E has property (P). Wegmann has proved [191, Theorem 5.5] that *if a normed linear space E has property* (P) *and G is an approximatively compact convex set in E, with property* (P), *then* $\mathscr{P}_G$ *is continuous*; also, Wegmann has given in $E = R^3$, an example [191] that in this result the assumption that G has property (P) cannot be omitted.

E. V. Oshman [135], [136] has proved that *every closed convex set G in a Banach space E is proximinal and with* $\mathscr{P}_G$ *u.s.c. if and only if E is reflexive,* $\mathscr{M}_f$ *is compact for each* $f \in E^* \setminus \{0\}$, *and the relations* $\{x_n\}_0^\infty \subset E, \{f_n\}_0^\infty \subset E^*, f_n(x_n) = \|x_n\| = \|f_n\| = 1, n = 0, 1, 2, \cdots, f_n \to f_0$ *for* $\sigma(E^*, E)$, $x_n \to x_0$ *weakly and* $\lim_{n\to\infty} \rho(x_n, G_{f_0} \cap G_{f_n}) = 0$, $\lim_{n\to\infty} \rho(x_0, G_{f_0} \cap G_{f_n}) = 0$, *where* $G_{f_n} = \{x \in E \mid f_n(x) = 1\}$, $n = 0, 1, 2, \cdots$, *imply* $\lim_{n\to\infty} x_n = x_0$. Furthermore, Oshman [135], [136] has characterized those separable Banach spaces E in which for every weakly boundedly compact closed convex set $G \subset E$, $\mathscr{P}_G$ is u.s.c. Similar results for $\mathscr{P}_G$ u.H.s.c. have been also obtained by Oshman [135], [136]. Also, Oshman has observed [139] that the isomorph of $E = l_R^2$, which does not have property (H), constructed in [166], has the property that *for every closed convex set* $G \subset E$, $\mathscr{P}_G$ *is u.H.s.c.* (*hence, in particular, for every convex Chebyshev set* $G \subset E$, π_G *is continuous*), *but for some closed hyperplane* $G_0 \subset E$, $\mathscr{P}_{G_0}$ *is not u.s.c.*; this space E of [166] is l_R^2 endowed with the equivalent norm $\|\cdot\|_2$ defined as the Minkowski functional of the set

$$S_2 = \{x = \{\xi_n\} \in l_R^2 \mid \|x\| \leqq 1, |\xi_1| \leqq \tfrac{1}{2}\}. \tag{5.12}$$

(e) Recently it was shown (see [177]) that the characterization theorems of the theory of best approximation can be generalized to convex optimization in locally convex spaces, using subdifferentials and directional derivatives. We recall that if φ is a continuous convex functional on a (real or complex) locally convex space E, (i.e., a continuous functional $E \to (-\infty, +\infty) = R$ such that $\varphi(\lambda x + (1 - \lambda)y) \leqq \lambda\varphi(x) + (1 - \lambda)\varphi(y)$ for all $x, y \in E$ and $0 \leqq \lambda \leqq 1$), and if G is a convex set in E, an element $g_0 \in G$ is called a *solution of the* (*continuous convex*) *program* $(G, \varphi|_G)$ if we have

$$\varphi(g_0) = \inf_{g \in G} \varphi(g); \tag{5.13}$$

we shall denote by $\mathscr{S}_G(\varphi)$ the set of all such elements g_0, i.e.,

$$\mathscr{S}_G(\varphi) = \{g_0 \in G \mid \varphi(g_0) = \inf_{g \in G} \varphi(g)\}. \tag{5.14}$$

Clearly, the problem of characterization of elements of best approximation of an element is a particular case of that of the characterization of solutions of a continuous convex program $(G, \varphi|_G)$ (i.e., of finding necessary and sufficient conditions in order that $g_0 \in \mathcal{S}_G(\varphi)$), by taking E = a normed linear space, G = a convex set in E, $x \in E \setminus \bar{G}$ and

$$\varphi(y) = \|x - y\|, \qquad y \in E. \tag{5.15}$$

We recall that if φ is a continuous convex functional on a locally convex space E, then for each $y_0 \in E$ the set

$$\partial\varphi(y_0) = \{f \in E^* | \operatorname{Re} f(y - y_0) \leqq \varphi(y) - \varphi(y_0)(y \in E)\} \tag{5.16}$$

is a nonvoid $\sigma(E^*, E)$-compact convex subset of E^* (see, for example, [82, p. 22]), called the *subdifferential* of φ at x_0 and that for each y_0, $y \in E$ the limit

$$\varphi'(y_0 ; y) = \lim_{t \to 0^+} \frac{\varphi(y_0 + ty) - \varphi(y_0)}{t} \tag{5.17}$$

exists (see, for example, [82, p. 16]) and is called the *directional derivative* of φ at y_0 in the direction y. We have the following characterizations of solutions of continuous convex programs [177].

THEOREM 5.3. *Let G be a convex set in a locally convex space E and let φ be a continuous convex functional on E and $g_0 \in G$. The following statements are equivalent;*

1°. $g_0 \in \mathcal{S}_G(\varphi)$.

2°. *We have*

$$\min_{f \in \mathcal{E}(\partial\varphi(g_0))} \operatorname{Re} f(g_0 - g) \leqq 0, \qquad g \in G. \tag{5.18}$$

3°. *We have*

$$\varphi'(g_0 ; g - g_0) \geqq 0, \qquad g \in G. \tag{5.19}$$

If G is a linear subspace of E, these statements are equivalent to the following:

4°. *We have*

$$0 \in \widetilde{\operatorname{co}} \{\mathcal{E}(\partial\varphi(g_0))|_G\}, \tag{5.20}$$

where $\widetilde{\operatorname{co}}$ stands for the $\sigma(G^*, G)$-closed convex hull.

From these characterizations one obtains as particular cases § 1, Theorems 1.2, 1.3 and 1.4 and extensions of Theorems 1.2 and 1.3 to best approximation by elements of convex sets G, by taking E to be a normed linear space and φ of the form (5.15), where $x \in E \setminus \bar{G}$, since it is easy to show that for this particular φ we have, for any $y_0 \in E$ [177],

$$\partial\varphi(y_0) = \{f \in E^* | f(y_0 - x) = \|y_0 - x\|, \|f\| = 1\}. \tag{5.21}$$

There also exists an extension of Theorem 5.1 (hence also of § 1, Theorem 1.1) to convex optimization, namely, a well-known theorem of Pshenichnii and Rockafellar (see, for example, [82, pp. 30–31]), which says that (for example, under the assumptions of Theorem 5.3) *we have $g_0 \in \mathcal{S}_G(\varphi)$ if and only if there exists an $f \in E^*$ such that*

$$f \in \partial\varphi(g_0), \tag{5.22}$$

$$\operatorname{Re} f(g_0 - g) \leqq 0, \qquad g \in G, \tag{5.23}$$

(actually, to get Theorem 5.1, one has to take $-f$ instead of f). In the particular case when G is a linear subspace of E, (5.23) reduces to $f \in G^{\perp}$ and thus we obtain a result of P. J. Laurent [118, p. 437], which is still an extension of § 1, Theorem 1.1.

For finite-dimensional convex sets G we have the following extension of Theorem 5.2 (hence also of § 1, Theorem 1.10).

THEOREM 5.4. *Let G be a convex set in a locally convex space E, such that* dim $[G] = n < \infty$ *(where $[G]$ is the closed linear subspace of E spanned by G) and let φ be a continuous convex functional on E and $g_0 \in G$. We have $g_0 \in \mathcal{S}_G(\varphi)$ if and only if there exist h extremal points $f_1, \cdots, f_h$ of $\partial\varphi(g_0)$, where $1 \leqq h \leqq n + 1$ if the scalars are real and $1 \leqq h \leqq 2n + 1$ if the scalars are complex, and h numbers $\lambda_1, \cdots, \lambda_h > 0$ with $\sum_{j=1}^{h} \lambda_j = 1$, such that*

$$\text{Re} \sum_{j=1}^{h} \lambda_j f_j(g_0 - g) \leqq 0, \qquad g \in G. \tag{5.24}$$

Theorem 5.4 was proved in [177]; in the particular case when G is a subspace of E, (5.24) reduces to $\sum_{j=1}^{h} \lambda_j f_j \in G^{\perp}$ and thus we obtain a result of P. J. Laurent [118, p. 438, Theorem 8.3.3], which is an extension of § 1, Theorem 1.10 (we note that in [118] only real scalars have been considered and different methods have been used).

We hope, in a continuation of [177], to show that much of the theory of best approximation can be extended to convex optimization in locally convex spaces. There exist only a few results in this direction. Thus, E. W. Cheney and A. A. Goldstein [40] have proved some existence theorems, for example, the following extension of § 2, Theorem 2.14, implication 5° ⇒ 1°: *If G is a closed convex set in a reflexive Banach space E, then every lower semi-continuous functional φ on E having bounded convex "level sets" $\{y \in E | \varphi(y) \leqq c\}$ attains its infimum on G.* The following extension of a weaker version of property (U) (see § 3, Corollary 3.1) has been introduced and studied by R. B. Holmes [80]: A linear subspace G of a real locally convex space E is said to have the *property* (φ-UHB), where φ is a given continuous convex functional on E, if each $\psi \in G^*$ satisfying $\psi \leqq \varphi|_G$ and $\{g \in G | \psi(g) = \varphi(g) \neq 0\} \neq \emptyset$, has at most one extension $f \in E^*$ satisfying $f \leqq \varphi$.

5.2. Best approximation by elements of N-parameter sets. One of the most important classical problems of best approximation by elements of nonlinear sets is that of best rational approximation in $E = C_R([a, b])$, raised by P. L. Chebyshev (see N. I. Ahiezer [1, Chap. II]), i.e., the problem of best approximation in $E = C_R([a, b])$ by elements of the set

$$G = R_{n,m} = \left\{ g \in E \,\middle|\, g(t) = z(t) \frac{\sum_{i=1}^{n} \alpha_i t^{i-1}}{\sum_{i=1}^{m} \beta_i t^{i-1}}, \sum_{i=1}^{m} \beta_i t^{i-1} \neq 0 \ (t \in [a, b]) \right\}, \tag{5.25}$$

where n, m are given positive integers and z is a given function in $E = C_R([a, b])$, such that $z(t) \geqq 0$ $(t \in [a, b])$; obviously, in the particular case when $m = 1$ and

$z(t) \equiv 1$, this problem reduces to that of best approximation by elements of the n-dimensional linear subspace $G_n = [1, t, \cdots, t^{n-1}]$ of $E = C_R([a, b])$. A slightly more general problem is that of best approximation in $E = C_R([a, b])$ by elements of the set

$$G = R_{G_1,G_2} = \left\{ g \in E \,\middle|\, g(t) = z(t)\frac{g_1(t)}{g_2(t)}, g_1 \in G_1, g_2 \in G_2, \right. \quad (5.26)$$
$$\left. g_2(t) \neq 0\ (t \in [a, b]) \right\},$$

where G_1, G_2 are given finite-dimensional linear subspaces of $E = C_R([a, b])$ and where z is as above. One can also replace the condition $g_2(t) \neq 0$ $(t \in [a, b])$ by weaker ones. A further generalization of the problem consists in replacing the interval $[a, b]$ by a compact space Q and the set R_{G_1,G_2} by an "N-parameter set", i.e., by a set of the form

$$G = \{g \in E = C_R(Q) | g = g(\alpha_1, \cdots, \alpha_N, q)(\{\alpha_1, \cdots, \alpha_N\} \in P, q \in Q)\}, \quad (5.27)$$

where P is a subset of a real N-dimensional Banach space ($N < \infty$), say B_N. The aim is to find classes of sets (5.27) such that the known results of the theory of rational approximation or of convex approximation (in particular, of linear approximation), e.g., the alternation theorem, the characterization theorem of Kolmogorov (see § 1, Theorem 1.9), uniqueness theorems, etc., remain valid for the sets G of these classes. For this purpose, there have been introduced, by various authors, "interpolating" N-parameter sets $G \subset E = C_R(Q)$, (i.e., having the property described in § 3, Definition 3.4, with $n = N$; they are also called "unisolvent" N-parameter sets), and then "locally unisolvent" and "asymptotically convex" N-parameter sets $G \subset E = C_R(Q)$, and, in an attempt to include also other important nonlinear sets G of approximating functions (e.g., of functions of the type $g(t) = \alpha_1\alpha_2^t + \alpha_3$), "varisolvent" N-parameter sets. The problem of necessary and sufficient conditions on (5.27) in order that a certain known theorem on rational approximation remain valid for (5.27) has been also studied; for example, local unisolvence is necessary and sufficient in order that the alternation theorem remain valid. The literature of these approximation problems in $E = C_R([a, b])$ and in $E = C_R(Q)$ (Q compact) is very vast. The reader may consult the monographs of N. I. Ahiezer [1], E. W. Cheney [38], G. Meinardus [127], J. R. Rice [151] and B. Brosowski [19], and the papers in the bibliographies of these monographs; for more recent results see, e.g., B. Brosowski and L. Wuytack [30] (on characterization and uniqueness), R. B. Barrar and H. L. Loeb (on the continuity of π_G) [6] and D. Braess [17] (on extensions of § 3, Theorem 3.11 from Chebyshev systems to varisolvent sets) and their references.

From the above it is clear that it would be important to develop a theory of best approximation by N-parameter sets G in a general normed space E, i.e., by sets

$$G = \{g = g(\alpha_1, \cdots, \alpha_N) \in E | \{\alpha_1, \cdots, \alpha_N\} \in P\}, \quad (5.28)$$

where P is a subset of a real N-dimensional Banach space B_N; naturally, such a theory would include, as particular cases, the above theories. This problem was first raised in [162, p. 137], and it is difficult even when $\dim E < \infty$. Two different approaches to this problem have been proposed by J. R. Rice [152], [153], [151] (see also [168, pp. 371–374]) and D. E. Wulbert [194], [195], [196] respectively. Both authors agree in pointing out the importance of the particular case when (5.28) is a manifold and obtain more results for this case; unfortunately, this does not include completely the rational approximation, since the set $G = R_{n,m}$ defined by (5.25) need not be a manifold in $E = C_R([a, b])$, even when $n = m = 2$. While the approach of J. R. Rice points out the importance of the concept of "curvature", and insists more on the case when $\dim E < \infty$, that of D. E. Wulbert emphasizes the utility of "boundedly connectedness". Among other results, D. E. Wulbert [196] has obtained a characterization of those n-dimensional C^1-submanifolds G of $E = C_R(Q)$ (Q compact) which are Chebyshev sets and satisfy a certain additional condition. We shall not enter here into more details. Let us only mention that often differentiability is used to linearize the problem and to draw from the known linear results for the nonlinear case, by observing that "local" best approximation (i.e., minimizing $\|x - g\|$ on a neighborhood of g_0) is equivalent to best approximation by elements of the "tangent" linear manifold to G passing through g_0; naturally, this contains best approximation by elements of linear subspaces G as a particular case, since it is known (see [168, p. 90]) that for linear subspaces G any element of local best approximation is already in $\mathscr{P}_G(x)$, i.e., is a "global" best approximation.

5.3. Generalizations. The problem of best approximation by elements of N-parameter sets admits further generalizations. We shall mention here two directions of such generalizations. Both of them have been considered first in $E = C_R(Q)$ and then in general normed linear spaces.

(a) The set P of parameters $\{\alpha_1, \cdots, \alpha_N\}$ in (5.27) can be replaced by a subset P of an infinite-dimensional normed linear space F; usually P is assumed to be open in F. For this case, assuming also Fréchet differentiability with respect to the parameter, G. Meinardus and D. Schwedt have given in $E = C_R(Q)$ a necessary condition for an element of best approximation (see [127, p. 140, Theorem 89], or [20, p. 28, Theorem 5]), which extends the necessity part of § 1, Theorem 1.9. In general this condition is not a sufficient one and the problem of characterization of the sets $G \subset E = C_R(Q)$ for which this condition is also sufficient, raised by B. Brosowski [20], has been solved recently by B. Brosowski and R. Wegmann [28].

In an arbitrary normed linear space E, P. J. Laurent [117] has given, under similar assumptions, necessary conditions for an element of best approximation which extend the necessity parts of § 1, Theorems 1.1 and 1.10.

(b) The set P of parameters can be completely omitted and one can consider the problem of finding classes of sets $G \subset E$ such that the results of the theory of N-parameter approximation or of convex approximation (in particular, of linear approximation) remain valid for the sets G of these classes. The notions of interpolating (=unisolvent), locally unisolvent, asymptotically convex and varisolvent sets G, mentioned in § 5.2, do not solve the problem in $E = C_R(Q)$, since they assume that G is an N-parameter set. B. Brosowski [19] has introduced the notion of a "regular" set G in $E = C_R(Q)$, which is independent of the notion of N-parameter set, and has proved that *for a set $G \subset C_R(Q)$ Kolmogorov's criterion* (§ 1, Theorem 1.9) *gives a necessary and sufficient condition in order to have $g_0 \in \mathscr{P}_G(x)$ if and only if G is a regular set* (we recall that, as shown in § 1, for any set G Kolmogorov's criterion gives a *sufficient* condition in order that $g_0 \in \mathscr{P}_G(x)$). A set $G \subset C_R(Q)$ is called [19] *regular*, if for every pair of elements $g, g_0 \in G$, every $\lambda > 0$ and every closed subset $A \subset Q$ such that $g(q) - g_0(q) \neq 0$ $(q \in A)$, there exists an element $g_\lambda \in G$ such that

$$\operatorname{sign}(g(q) - g_0(q)) = \operatorname{sign}(g_\lambda(q) - g_0(q)), \qquad q \in A, \tag{5.29}$$

$$\|g_\lambda - g_0\| < \lambda. \tag{5.30}$$

From the above result on Kolmogorov's criterion it follows, in particular, that every convex set G, every set $G = R_{G_1,G_2}$ of the form (5.26), every varisolvent N-parameter set G and every asymptotically convex set G in $E = C_R(Q)$ is regular (naturally, this can be deduced also directly from the definitions; see [19]). Some other problems of best approximation in $E = C_R(Q)$ by elements of regular sets G (for example, uniqueness), have been also studied by B. Brosowski [19].

In arbitrary normed linear spaces the similar problem (of finding classes of sets $G \subset E$ such that the results of the theory of N-parameter approximation or of convex approximation carry over to the sets G of these classes) has also been studied. One of the important notions in this direction has turned out to be that of a sun. A set G in a normed linear space E is called a *sun*, if $g_0 \in \mathscr{P}_G(x)$ implies

$$g_0 \in \mathscr{P}_G(\lambda x + (1 - \lambda)g_0), \qquad 1 \leqq \lambda < \infty, \tag{5.31}$$

that is, if whenever $g_0 \in \mathscr{P}_G(x)$, then g_0 is also an element of best approximation to every point on the ray from g_0 through x; the term "sun" is apparently due to N. V. Efimov and S. B. Stechkin [59], but such sets have been considered also earlier (see, e.g., V. Klee [103]). It is easy to see that *every convex set G is a sun*; indeed, if G is convex and $g_0 \in \mathscr{P}_G(x)$, $1 \leqq \lambda < \infty$, then

$$\begin{aligned}\|\lambda x + (1 - \lambda)g_0 - g_0\| &= \lambda\|x - g_0\| \\ &\leqq \lambda\left\|x - \left(\frac{1}{\lambda}g + \left(1 - \frac{1}{\lambda}\right)g_0\right)\right\| = \|\lambda x + (1 - \lambda)g_0 - g\|, \quad g \in G.\end{aligned}$$

V. Klee [103, Theorem 2] and later, independently, N. V. Efimov and S. B. Stechkin (see [168, pp. 366–368]) have proved the following partial converse (a shorter proof, using Theorem 5.5 below, has been given by D. Amir and F. Deutsch [2]): *In a smooth normed linear space E, every proximinal sun is convex.* Also, B. Brosowski and R. Wegmann [28] have proved (a short direct proof has been given by Brosowski and F. Deutsch [23]) that *if G is a proximinal sun in a normed linear space E, then*

$$\operatorname{co} \mathscr{P}_G(x) \subset \operatorname{Fr} S(x, \rho(x, G)), \qquad x \in E; \tag{5.32}$$

hence, in particular, *in a strictly convex space E every proximinal sun is a Chebyshev set* (naturally, for smooth strictly convex spaces E this also follows from the above result of Klee-Efimov-Stechkin). B. Brosowski has proved [21] that a *local element of best approximation by elements of a sun is already global.*

We have seen above that best approximation by elements of suns includes as a particular case best approximation by elements of convex sets and we shall see (by Theorem 5.5 below) that it also includes best approximation in $E = C_R([a, b])$ by elements of the N-parameter set $G = R_{n,m}$ defined by (5.25); however, it does not include as a particular case the same problem in the smooth spaces $E = L^p([a, b], v)$, where v is the Lebesgue measure and where $1 < p < \infty$, since $G = R_{n,m}$ is proximinal but not convex when $m \geqq 2$ and hence, by the above result of Klee-Efimov-Stechkin, it is not a sun in E.

We have seen in § 1 that for any set $G \subset E$ Kolmogorov's criterion gives a sufficient condition in order that $g_0 \in \mathscr{P}_G(x)$. Following B. Brosowski and R. Wegmann [28], a set G in a normed linear space E is called a *Kolmogorov set*, if this condition is also necessary, i.e., if $g_0 \in \mathscr{P}_G(x)$ implies

$$(5.33)\qquad \min_{f \in \mathscr{E}(S_{E^*}) \cap M_{x-g_0}} \operatorname{Re} f(g - g_0) \leqq 0, \qquad g \in G,$$

where $M_{x-g_0} = \{f \in E^* | \, \|f\| = 1, f(x - g_0) = \|x - g_0\|\}$. Finally, the notion of a regular set G in $C_R(Q)$, given above, admits the following natural generalization, due to B. Brosowski and R. Wegmann [28] : a set G in a normed linear space E is called *regular*, if for every pair of elements $g_0, g \in G$, every $\lambda > 0$, $x \in E \diagdown G$ and every $\sigma_{\mathscr{E}}$-closed set $A \subset \mathscr{E}(S_{E^*})$ containing M_{x-g_0} and such that $\operatorname{Re} f(g - g_0) > 0$ for all $f \in A$, there exists an element $g_\lambda \in G$ such that

$$(5.34)\qquad \operatorname{Re} f(g_\lambda - g) > \operatorname{Re} f(x - g_0) - \|x - g_0\|, \qquad f \in A,$$

$$(5.35)\qquad \|g_\lambda - g_0\| < \lambda ;$$

here $\sigma_{\mathscr{E}}$ denotes the topology on $\mathscr{E}(S_{E^*})$ induced by $\sigma(E^*, E)$.

THEOREM 5.5. *For a set G in a normed linear space E the following statements are equivalent*:

1°. *G is a Kolmogorov set.*

2°. *We have*

$$(5.36)\qquad G \cap \operatorname{Int} S(\lambda x + (1 - \lambda) g_0, \lambda \|x - g_0\|) = \varnothing, \qquad g_0 \in \mathscr{P}_G(x), 0 < \lambda < \infty .$$

3°. *G is a sun.*

4°. *G is a regular set.*

The equivalence 1° ⇔ 3° has been proved by B. Brosowski [21] and another proof, via 1° ⇔ 2° ⇔ 3°, has been given by D. Amir and F. Deutsch [2]; the equivalence 1° ⇔ 4° is due to B. Brosowski and R. Wegmann [28].

The suns defined above by (5.31) are also called [106] *α-suns*. There exist also other notions of suns. Thus, a set G in a normed linear space E is called [106] a *β-sun*, if for every $x \in E$ with $\mathscr{P}_G(x) \neq \varnothing$ and every $g_0 \in \mathscr{P}_G(x)$ we have

$$\bigcap_{1 \leqq \lambda < \infty} \mathscr{P}_G(\lambda x + (1 - \lambda)g_0) \neq \varnothing. \tag{5.37}$$

L. P. Vlasov has proved (see [21]) that *in a smooth normed linear space every proximinal β-sun is convex.* B. Brosowski has shown [21] that in a strictly convex space E every proximinal *β-sun is a Chebyshev set.* Since obviously every α-sun is a β-sun, these results imply the abovementioned results of Klee-Efimov-Stechkin and Brosowski-Wegmann (the statement after (5.32)) respectively.

THEOREM 5.6. *For a set G in a normed linear space E the following statements are equivalent*:

1°. $\mathscr{P}_G(x) \neq \varnothing$ *implies the existence of an element* $g_0 \in \mathscr{P}_G(x)$ *satisfying* (5.33).

2°. *G is a β-sun.*

If G is proximinal and if $\mathscr{P}_G(x)$ *is compact for every* $x \in E$, *these statements are equivalent to the following*:

3°. *For every* $x \in E$ *and* $r > 0$ *the set-valued mapping* $\Lambda_{x,r}: E \to 2^E$ *defined by*

$$\Lambda_{x,r}(y) = x + \frac{r}{r + \rho(x, G)}(y - \mathscr{P}_G(y)), \qquad y \in E, \tag{5.38}$$

has a fixed point (i.e., a point $y_0 \in E$ such that $y_0 \in \Lambda_{x,r}(y_0)$).

The equivalence 1° ⇔ 2° has been proved by B. Brosowski [21] and the equivalence 1° ⇔ 3° by B. Brosowski, K. H. Hoffmann, E. Schäfer and H. Weber [26]. Some other characterizations of the above classes of sets in terms of fixed points of set-valued mappings have been given by B. Brosowski [22]. A localized Kolmogorov type criterion, in terms of cones of displacement, has been also given by B. Brosowski [21] and the sets $G \subset E$ for which this criterion is necessary and sufficient in order that $g_0 \in \mathscr{P}_G(x)$ have been characterized by B. Brosowski and R. Wegmann [28].

D. Amir and F. Deutsch have introduced [2] the following generalization of the notion of a sun: A set G in a normed linear space E is called a *moon*, if the relations $g_0 \in G$, $K(g_0, x) \cap G \neq \varnothing$ imply $g_0 \in \overline{K(g_0, x) \cap G}$, where $K(g_0, x)$ is the convex cone with vertex g_0 defined by

$$K(g_0, x) = \left\{ y \in E \,\middle|\, \min_{f \in M_{x - g_0}} \operatorname{Re} f(y - g_0) > 0 \right\}; \tag{5.39}$$

it has been proved in [2] that

$$K(g_0, x) = \bigcup_{\lambda > 0} \operatorname{Int} S(\lambda x + (1 - \lambda)g_0, \ \ \lambda\|x - g_0\|), \tag{5.40}$$

hence condition 2° of Theorem 5.5 can be also written in the form $G \cap K(g_0, x) = \varnothing$ $(g_0 \in \mathscr{P}_G(x))$. D. Amir and F. Deutsch have shown [2] that in the above definition of a moon one can replace $g_0 \in G$ by $g_0 \in \mathscr{P}_G(x)$; hence, in particular, *every sun is a moon.* Also [2], *if E is strictly convex, then* Fr S_E *is a moon* (*but not a sun*); *if* dim $E = 2$, *the converse is also true.* However, there are some classes of spaces in which every moon is a sun. A normed linear space E is called [2] *strongly nonlunar*, if for each $y_0 \in \operatorname{Fr} S_E$ and each $z \in K(y_0, 0)$ there exists an $x \in \operatorname{Int} S_E$ with $z \in K(y_0, x)$, such that $y_0 \notin \overline{K(g_0, x) \cap \operatorname{Fr} S_E}$. D. Amir and F. Deutsch have proved [2] that *in a strongly nonlunar space E a set $G \subset E$ is a moon if and only if it is a sun* and that the following concrete spaces are strongly nonlunar: $C_0(T)$, where T is locally compact, $l^1(S)$, where S is an arbitrary set, and every finite-dimensional polyhedral space. Some more classes of strongly nonlunar spaces have been given, in terms of a maximum property of the "peak set mapping" $x \to M_x = \{f \in E^* | \|f\| = 1, f(x) = \|x\|\}$, by R. Wegmann [191]. Amir and Deutsch have also raised [2] the problem whether the converse of the above result is true, i.e., *whether a space in which every moon is a sun must be strongly nonlunar.* The following partial answer has been proved by B. Brosowski and F. Deutsch [24, Theorem 2.6] : *If every moon G in a normed linear space E is sun, then for each $y_0 \in \mathrm{Fr}\, S_E$ there exists an $x \in \mathrm{Int}\, S_E$ such that $y_0 \notin \overline{K(g_0, x) \cap \mathrm{Fr}\, S_E}$.* Also B. Brosowski and F. Deutsch have proved [24, Theorem 2.8] that *if for a set G every local element of best approximation by the elements of G is already global, then G is a moon.*

The above notions can be localized (introducing solar, lunar, strongly nonlunar points, etc.; see, e.g., [21], [2]).

For other related results and applications to best approximation we refer the reader to the abovementioned papers.

5.4. Best approximation by elements of arbitrary sets.

(a) We have the following characterization of elements of best approximation in terms of fixed points of a set-valued mapping, given by O. Brandt [18] and B. Brosowski [22].

PROPOSITION 5.1. *Let E be a normed linear space, G an arbitrary set in E, $x \in E \setminus G$ and $g_0 \in G$. We have $g_0 \in \mathscr{P}_G(x)$ if and only if g_0 is a fixed point of the mapping $\mathscr{B}_x : G \to 2^G$ defined by*

$$\mathscr{B}_x(g) = G \cap S\left(x, \frac{\rho(x, G) + \|x - g\|}{2}\right), \qquad g \in G, \tag{5.41}$$

(i.e., $g_0 \in \mathscr{B}_x(g)$). Moreover, in this case we have

$$\mathscr{B}_x(g_0) \subset \mathscr{P}_G(x). \tag{5.42}$$

Indeed, we have $g_0 \in \mathscr{B}_x(g_0)$ if and only if $g_0 \in G$ and $\|x - g_0\| \leqq (\rho(x, G) + \|x - g_0\|)/2$, which is clearly equivalent to $g_0 \in \mathscr{P}_G(x)$. Also, if $g \in \mathscr{B}_x(g_0)$, where $g_0 \in \mathscr{P}_G(x)$, then $g \in G$ and $\|x - g\| \leqq (\rho(x, G) + \|x - g_0\|)/2 = \rho(x, G)$, whence $g \in \mathscr{P}_G(x)$, which completes the proof.

(b) There exist some useful sufficient conditions for a set G in a normed linear (or, more generally, in a metric) space E to be proximinal. In [168, pp. 383–384] it was proved

THEOREM 5.7. *If G is a set in a metric space E, for which there exists a topology τ on E* (not necessarily comparable with the initial topology of E) *such that for every $x \in E$ and $r > 0$ the intersection $G \cap S(x, r)$ is τ-compact, then G is proximinal.*

In particular, from this theorem one obtains the following extension of § 2, Theorem 2.9(i), stated by R. R. Phelps [145] for linear subspaces Γ, by E. W. Cheney and A. A. Goldstein [40, Theorem 12] for existence of elements of minimal norm in Γ, and by J. Blatter and E. W. Cheney [14] in the general case.

COROLLARY 5.1. *If Γ is a $\sigma(E^*, E)$-closed set in the conjugate space E^* of a normed linear space E, then Γ is proximinal.*

From Corollary 5.1 it follows, in particular, that *every weakly closed set G in a reflexive Banach space E is proximinal.* We shall see in part (e) below that Theorem 5.7 and Corollary 5.1 are useful for applications to best approximation of operators by elements of sets of continuous linear operators.

N. V. Efimov and S. B. Stechkin have proved (see [168, p. 382]) that *every approximatively compact (hence, in particular, every boundedly compact closed) set G in a metric space E is proximinal.*

Another problem on existence of elements of best approximation is that of finding the Banach spaces E with the property *that for every closed set $G \subset E$ the set*

$$\mathscr{D}(\pi_G) = \{x \in E | \mathscr{P}_G(x) \neq \emptyset\} \tag{5.43}$$

is dense in E. S. B. Stechkin [179] and M. Edelstein [56] have proved that *every uniformly convex Banach space E has this property.* This result has been slightly extended by D. E. Wulbert [194] who has proved that *every Banach space E "with property* (2R)" *also has the above property.* We recall (see, for example, [48]) that a Banach space E is said to have *property* (2R) if every sequence $\{x_n\} \subset S_E$ such that $\lim_{n,m \to \infty} \|x_n + x_m\| = 2$ is a Cauchy sequence (and hence convergent); clearly, every uniformly convex space has property (2R) but the converse is not true. Since every space with property (2R) (and hence every uniformly convex space) is strictly convex, it is natural to ask whether there exist non-strictly convex spaces E with the above property (i.e., such that for every closed set $G \subset E$ the set $\mathscr{D}(\pi_G)$ is dense in E). D. E. Wulbert [194] has given an affirmative answer, by proving that *every uniformly smooth Banach space E with property* (H) (see § 4.2) *also has the above property.* We recall (see, for example, [48]) that E is called *uniformly smooth* if for every $\eta > 0$ there exists an $\varepsilon = \varepsilon(\eta)$ such that the relation $\|x - y\| \leqq \varepsilon$ implies $\|x\| + \|y\| \leqq (1 + \eta)\|x + y\|$; D. E. Wulbert [194] has shown that there exist uniformly smooth spaces with property (H) which are not strictly convex.

By the remark made at the end of § 2 (on very nonproximinal subspaces) *a Banach space E with the above property must be reflexive.* D. E. Wulbert [194] has raised the problem whether the converse is true, that is:

Problem 5.1. Does there exist a reflexive Banach space E containing a closed set G such that $\mathscr{D}(\pi_G)$ is not dense in E?

Some other problems related to existence of elements of best approximation are concerned with very non-proximinal sets (see § 2, Definition 2.2). M. Edelstein [57] has proved that *in a separable conjugate space E^* no closed bounded set Γ is very non-proximinal.* He has also shown [57] that *in the separable space $E = c_0$* (which is not isomorphic to any conjugate Banach space) *there do exist bounded very non-proximinal sets.*

V. Klee (see [168, p. 371]) has considered the problem of characterization of the classes $\mathscr{N}_i$, $i = 0, 1, 2, 3, 4$, of all normed linear spaces E which contain a very non-proximinal set G having respectively the following properties: (0) no additional property; (1) G is convex; (2) G is bounded and convex; (3) $E \setminus G$ is convex; (4) $E \setminus G$ is bounded and convex. V. Klee has made the following remarks: (i) $\mathscr{N}_1$ is the class of all nonreflexive spaces; (ii) $\mathcal{N}_3 \supset \mathcal{N}_1$; (iii) $\mathcal{N}_2 \neq \emptyset$, since it contains a noncomplete space E; (iv) $\mathcal{N}_4 \neq \emptyset$; (v) it is possible that $\mathcal{N}_4$ (whence also $\mathcal{N}_3, \mathcal{N}_0$) coincides with the class of all normed linear spaces. In connection with (iii), M. Edelstein and A. C. Thompson [58] have shown that we also have $c_0 \in \mathcal{N}_2$, so $\mathcal{N}_2$ contains also a Banach space.

(c) For an arbitrary set G in a normed linear space E, S. B. Stechkin (see [168, p. 375]) has studied the set

$$\mathcal{U}_G = \{x \in E | g_1, g_2 \in \mathcal{P}_G(x) \Rightarrow g_1 = g_2\}, \tag{5.44}$$

i.e., the set of all elements $x \in E$ which have *at most* one element of best approximation in G, and has obtained, among other results, the following "constructive characterization" of strictly convex spaces.

THEOREM 5.8. *A Banach space E has the property that for every set $G \subset E$ the set $\mathcal{U}_G$ is dense in E if and only if E is strictly convex.*

B. Brosowski and F. Deutsch have proved [23, Theorem 3.6] the following partial extension of the sufficiency part of Theorem 5.8: *If E is a normed linear space such that every convex subset of* Fr S_E *is finite-dimensional and if G is a proximinal set in E, then the set*

$$\mathcal{U}_G^{(1)} = \{x \in E | \operatorname{co} \mathcal{P}_G(x) \subset \operatorname{Fr} S(x, \rho(x, G))\} \tag{5.45}$$

is dense in E. Obviously, we always have $\mathcal{U}_G \subset \mathcal{U}_G^{(1)}$ and, if E is strictly convex, then $\mathcal{U}_G = \mathcal{U}_G^{(1)}$ (since then every convex subset of Fr $S(x, \rho(x, G))$ is either void or a singleton), hence the above result gives again the sufficiency part of Theorem 5.8 for proximinal sets G. Also, Brosowski and Deutsch have given [23] an example (in $E = L^\infty([0, 1], \nu)$, where ν is the Lebesgue measure) that in general the assumption of finite dimensionality of every convex subset of Fr S_E cannot be omitted; however, they have proved [23, Theorem 3.10] that this can be done when E is smooth, i.e., that *if G is a proximinal set in a smooth normed linear space E then $\mathcal{U}_G^{(1)}$ is dense in E.*

Furthermore, S. B. Stechkin has proved [179] that *if E is a strictly convex Banach space, then for every boundedly compact set $G \subset E$ the set $\mathcal{U}_G$ is of the second category in E.* However, it is not known whether this also holds for every set $G \subset E$; it is also unknown whether from the fact that for every *compact* $G \subset E$ the set $\mathcal{U}_G$ is either dense in E or of the second category in E it follows that E is strictly convex. Finally, we mention the following results of Stechkin [179]: *if E is a locally uniformly convex Banach space, then for every $G \subset E$ the set $\mathcal{U}_G$ is of the second category and if E is a uniformly convex Banach space, then for every closed set G in E the set $\mathcal{D}(\pi_G) \cap \mathcal{U}_G$ is of the second category.* However, it is not known whether the second result remains valid also for locally uniformly convex spaces.

We have seen in § 5.3 (b) that in a strictly convex space E every proximinal sun is a Chebyshev set. Conversely, it is natural to ask (see, e.g., [23], [24]):

Problem 5.2. In a normed linear space E, is every Chebyshev set $G \subset E$ a sun?

We shall see below (Theorem 5.9 and the remarks made after it) that in some important cases (in particular, for dim $E < \infty$), the answer is affirmative.

Let us mention now a famous classical problem, which is a particular case of Problem 5.2 (since we have seen in § 5.3 (b) that in a smooth space every proximinal sun is convex), namely, the *problem of convexity of Chebyshev sets*. We have seen in § 5.1 that a Banach space E has the property that every closed convex set $G \subset E$ is a Chebyshev set if and only if E is reflexive and strictly convex. It is natural to ask what are the Banach spaces E in which the converse property holds, i.e., in which every Chebyshev set $G \subset E$ is convex. This problem has been solved only for 2- and 3-dimensional spaces E (see for example [168, p. 364]), namely, for $\dim E = 2$, E has this property if and only if it is smooth, and for $\dim E = 3$, if and only if every exposed point of S_E (see § 3.2) admits a *unique* maximal functional of norm 1. For Banach spaces E of finite dimension $m \geqq 4$ it is only known that the smoothness of E is a sufficient but not necessary condition for the convexity of all Chebyshev sets $G \subset E$. For infinite-dimensional Banach spaces E the problem is considerably more difficult, even the answer to the following problem being unknown.

Problem 5.3. In a Hilbert space H, is every Chebyshev set necessarily convex?

V. Klee has conjectured that the answer is negative and has proved (see [168, p. 370]) that *in every infinite-dimensional Hilbert space H there exist nonconvex closed semi-Chebyshev sets G*, even with $H \setminus G$ nonvoid, bounded and convex. On the other hand, much work has been done towards a positive answer. By an ingenious application of Schauder's fixed-point theorem, L. P. Vlasov has proved (see [168, p. 365])

THEOREM 5.9. *In an arbitrary Banach space E every boundedly compact Chebyshev set G is a sun (and hence, if E is smooth, G is convex).*

The assumption of boundedly compactness of G in this result was weakened by N. V. Efimov and S. B. Stechkin and others (see [168, pp. 368–369]), under additional restrictions on the space E. For example, we have the following infinite-dimensional characterizations of closed convex sets in terms of the Chebyshev property (see [168, p. 369]).

THEOREM 5.10.. *For a set G in a smooth uniformly convex Banach space E the following statements are equivalent*:

1°. *G is convex and closed.*

2°. *G is a weakly closed Chebyshev set.*

3°. *G is an approximatively compact Chebyshev set.*

More generally (see [60] and [183]), *in a locally uniformly convex Banach space E, every approximatively compact Chebyshev set is a sun (and hence, if E is smooth, then G is convex).*

We have seen in § 5.3 (b) that in the spaces $E = L^p([a, b], v)$, where v is the Lebesgue measure and where $1 < p < \infty$, the set $G = R_{n,m}$ defined by (5.25) is not a sun when $m \geqq 2$; hence, since G is weakly closed and approximatively compact, from Theorem 5.10 it follows that $G = R_{n,m}$ is not a Chebyshev set in E when $m \geqq 2$ (in contrast with the situation for $E = C_R([a, b])$, where $G = R_{n,m}$ is a Chebyshev set, by a classical theorem of P. L. Chebyshev). For a generalization to the sets $R = R_{G_1, G_2}$ defined by (5.26), on compact spaces, see J. Blatter [11].

L. P. Vlasov has proved [183, Theorem 1] that *if* $G \neq E$ *is an approximatively compact Chebyshev set in a Banach space E, then* $E \setminus G$ *is unbounded.* However, [4], *if there exists a nonconvex Chebyshev set in a Hilbert space H, then there also exists in H a nonconvex Chebyshev set G such that* $H \setminus G$ *is bounded and convex.*

L. P. Vlasov has proved [186, Theorem 7] that *in a uniformly convex space E, every locally compact Chebyshev set G (i.e., such that for each* $g \in G$ *there exists a* $\delta > 0$ *for which* $G \cap S(g, \delta)$ *is compact) is a sun (and hence, if E is smooth, then G is convex)*; this improves some results of D. E. Wulbert [192].

An important further step was the idea of V. Klee of imposing continuity conditions on the metric projection π_G onto G rather than imposing conditions directly on the Chebyshev set G; in this way, for all classes of Chebyshev sets G for which π_G has the required continuity properties, it follows that the sets G in those classes are convex. (Note also that, conversely, *if G is a closed convex set in a smooth uniformly convex Banach space E, then* by Theorem 5.10 *G is an approximatively compact Chebyshev set* and hence, by § 4.7 (f), the result after Theorem 4.24, π_G *is continuous.*) L. P. Vlasov [188] has proved

THEOREM 5.11. *If E is a Banach space such that the conjugate space* E^* *is strictly convex (hence E is a smooth Banach space), then every Chebyshev set* $G \subset E$ *with continuous metric projection* π_G *is convex.*

For Hilbert spaces E. Asplund [4] has shown that it is sufficient here to assume that π_G is continuous from the norm topology to the weak topology. Also, E. Asplund [4] has proved

THEOREM 5.12. *If G is a Chebyshev set in a Hilbert space H such that every closed half-space intersects G in a proximinal set, then G is convex.*

These two theorems contain as particular cases the previously known results, since, for example, every boundedly compact Chebyshev set G satisfies the above hypotheses. Let us note that the arguments of E. Asplund [4] lean heavily on the tools of the theory of convex functionals.

D. E. Wulbert has proved [192] that *in a normed linear space E, every Chebyshev set* $G \subset E$ *with continuous metric projection* π_G *is boundedly connected* (i.e., intersects every open cell in a connected set) and hence connected. Extending some results of Wulbert, L. P. Vlasov [186, Theorem 5] has proved that *in a Banach space E, every locally compact Chebyshev set G with continuous metric projection* π_G *is a sun (and hence, if E is smooth, then G is convex).*

Various generalizations of convexity and of locally uniformly convex spaces, with applications to Chebyshev sets, and characterizations, in some spaces, of all Chebyshev sets G with continuous metric projection π_G, have been given by L. P. Vlasov [182]–[189] and E. V. Oshman [137], [138].

(d) Let us mention now some extensions of the above results to more general (not necessarily Chebyshev) sets G, which state that under certain assumptions on G and on the sets $\mathcal{P}_G(x)$ for each $x \in E$, or on the set-valued metric projection $\mathcal{P}_G$, the set G is convex or has some other more general properties.

L. P. Vlasov has proved the following extension of Theorem 5.9 (see [168, p. 368]): *In a smooth Banach space E, if G is a boundedly compact closed set such that* $\mathcal{P}_G(x)$ *is convex for each* $x \in E$, *then G is convex.* Also, Vlasov has proved

[185, Theorem 4] that in a reflexive space E, if G is a proximinal set such that (i) *$\mathcal{P}_G(x)$ is compact and convex for each $x \in E$,* (ii) *$\mathcal{P}_G$ is u.H.s.c., and* (iii) *$\varphi(x) = \rho(x, G)$ is weakly u.s.c., then G is a sun* (*and hence, if E is smooth, then G is convex*). Here (iii) can be also replaced by the condition that $E \setminus G$ be convex [185, Theorem 5] or by the condition that E be "compactly locally uniformly rotund", i.e., that the relations $\|x_n\| = \|x_0\| = 1$, $n = 1, 2, \cdots$, $\lim_{n\to\infty} \|x_n - x_0\| = 2$ imply the existence of a limit point for $\{x_n\}$ [185, Theorem 6]; if E is, in addition, smooth, then (iii) can be omitted [185, Theorem 7].

W. Pollul [148, p. 56, Theorem 6] has proved

THEOREM 5.13. *If G is a proximinal set in a normed linear space E and $\mathcal{P}_G$ is l.s.c. at some $x_0 \in E \setminus G$, then*

$$\text{co}\, \mathcal{P}_G(x_0) \subset \text{Fr}\, S(x_0, \rho(x_0, G)) \tag{5.46}$$

(*and hence, if G is proximinal and $\mathcal{P}_G$ is l.s.c., then we have* (5.32)).

Theorem 5.13 improves a result of Blatter, Morris and Wulbert [15, Theorem 15], in which it was assumed that $\mathcal{P}_G$ is continuous. From Theorem 5.13 it follows ([148, p. 56, Corollary 1] and when $\mathcal{P}_G$ is continuous [15, Corollary 16]) that *if G is a proximinal set in a strictly convex normed linear space E and $\mathcal{P}_G$ is l.s.c. at x_0, then $\mathcal{P}_G(x_0)$ is a singleton* (*and hence, if $\mathcal{P}_G$ is l.s.c., then G is a Chebyshev set*). J. Blatter has observed [12] that this also follows immediately from the sufficiency part of Theorem 5.8; indeed, since $\mathcal{U}_G$ is dense in E, one can choose a sequence $\{x_n\} \subset \mathcal{U}_G$ such that $\lim_{n\to\infty} x_n = x_0$, whence, since $\mathcal{P}_G$ is l.s.c. at x_0, for any $g_0 \in \mathcal{P}_G(x_0)$ we must have $\lim_{n\to\infty} \pi_G(g_n) = g_0$, so $\mathcal{P}_G(x_0)$ is singleton. From Theorem 5.13 it also follows ([148, p. 56, Corollary 2] and [15, Corollary 17]) that *in a smooth strictly convex Banach space E a boundedly compact set G is a Chebyshev set* (*or, equivalently, convex—see Theorem 5.9 above*) *if and only if $\mathcal{P}_G$ is continuous.*

J. Blatter has proved [12, Theorem 7] that *if G is a proximinal set in a smooth locally uniformly convex* (hence strictly convex) *Banach space E and $\mathcal{P}_G$ is l.s.c., then G is convex*; this is an extension of the result given after Theorem 5.10 above. Hence, as J. Blatter has observed, it follows [12, Theorem 8] that *for a set G in a smooth locally uniformly convex reflexive Banach space E the following statements are equivalent*: 1°. *G is an approximatively compact Chebyshev set.* 2°. *G is convex and closed.* 3°. *G is a Chebyshev set with π_G continuous.* 4°. *G is a proximinal set with $\mathcal{P}_G$ l.s.c.*

Another result of Vlasov, mentioned after Theorem 5.10 above, admits the following extension [185, Theorem 2]: *If $G \neq E$ is a proximinal set in a Banach space E, such that* (i) *$\mathcal{P}_G(x)$ is compact and convex for each $x \in E$ and* (ii) *$\mathcal{P}_G$ is u.H.s.c., then $E \setminus G$ is unbounded.* On the other hand, J. Blatter has proved [12, Theorem 6] that *if $G \neq E$ is a proximinal set in a Banach space E, such that $\mathcal{P}_G$ is l.s.c., then $E \setminus G$ is unbounded.*

W. Pollul has proved [148, p. 56, Theorem 2] the following extension of Wulbert's theorem given after Theorem 5.12 above: *If G is a proximinal set in a normed linear space E, such that $\mathcal{P}_G$ is continuous, then G is boundedly connected and $\mathcal{P}_G(x)$ is connected for each $x \in E$.* This improves a result of Blatter, Morris and Wulbert

[15, Theorem 14], in which it was assumed that $\mathscr{P}_G(x)$ is compact for each $x \in E$.

We have seen in § 4.7, that if G is a linear subspace of a normed linear space E, such that $\mathscr{P}_G$ is u.s.c., then $\mathscr{P}_G(x)$ is compact for each $x \in E$. B. Brosowski and F. Deutsch have observed that the proof of this result, given in [174], uses actually only that $\mathscr{P}_G$ is u.s.c. along certain rays. This has led them to introduce [23] for the metric projection $\mathscr{P}_G$ the following notions of "radial" semi-continuity, which are weaker than the corresponding notions of semi-continuity and Hausdorff semi-continuity.

DEFINITION 5.1. Let G be a set in a normed linear space E and let $x_0 \in E$. The metric projection $\mathscr{P}_G : E \to 2^G$ is called:

(i) *outer radially upper semi-continuous* (OR u.s.c.) *at* x_0, if the relations $g_0 \in \mathscr{P}_G(x_0)$, $\{x_n\} \subset \{g_0 + \lambda(x_0 - g_0) | 1 \leqq \lambda < \infty\}$ and $\lim_{n\to\infty} x_n = x_0$ imply $\lim_{n\to\infty} \sup_{g \in \mathscr{P}_G(x_n)} \rho(g, \mathscr{P}_G(x_0)) = 0$;

(ii) *outer radially lower semi-continuous* (OR l.s.c.) *at* x_0, respectively *inner radially lower semi-continuous* (IR l.s.c.) *at* x_0, if the relations $g_0, g_1 \in \mathscr{P}_G(x_0)$, $\{x_n\} \subset \{g_0 + \lambda(x_0 - g_0) | 1 \leqq \lambda < \infty\}$, respectively $\{x_n\} \subset \{g_0 + \lambda(x_0 - g_0) | 0 \leqq \lambda \leqq 1\}$, and $\lim_{n\to\infty} x_n = x_0$ imply $\lim_{n\to\infty} \rho(g_1, \mathscr{P}_G(x_n)) = 0$;

(iii) τ *semi-continuous* (τ.s.c.), where $\tau =$ ORu., ORl. or IRl., when $\mathscr{P}_G$ is τ.s.c. at each $x_0 \in E$.

As mentioned above, B. Brosowski and F. Deutsch have observed that the proof of [174, Theorem 1], yields also the following more general result [23, Theorem 4.4]: *If G is a proximinal set in a normed linear space E, such that $\mathscr{P}_G$ is OR u.s.c. and that $\mathscr{P}_G(x)$ is convex for each $x \in E$, then $\mathscr{P}_G(x)$ is compact for each $x \in E$.* Since OR upper semi-continuity is weaker than the usual upper semi-continuity, one can also prove the following partial converse [23, Theorem 4.5]: *If G is a sun in a normed linear space E, such that $\mathscr{P}_G(x)$ is compact for each $x \in E$, then $\mathscr{P}_G$ is OR u.s.c.* Combining these two results it follows [23, Corollary 4.7] that *if G is a proximinal convex set, $\mathscr{P}_G$ is OR u.s.c. if and only if $\mathscr{P}_G(x)$ is compact for each $x \in E$.* Furthermore, combining this latter result with § 4, Theorem 4.25, equivalence $1° \Leftrightarrow 4°$, it follows [23, Corollary 4.8] that *for a closed hyperplane G, $\mathscr{P}_G$ is OR u.s.c. if and only if it is u.s.c.*

B. Brosowski and F. Deutsch have proved [24, Theorem 2.10] that if *G is a sun in a normed linear space E, then $\mathscr{P}_G$ is OR l.s.c.* and [24, Theorem 2.11] that *if $\mathscr{P}_G$ is OR l.s.c., then every local element of best approximation by the elements of G is already global* (and hence, as mentioned at the end of § 5.3, G is a moon). Also, they have proved [23, Theorem 3.3] that *if $\mathscr{P}_G(x_0)$ is convex, then $\mathscr{P}_G$ is IR l.s.c. at x_0.* Hence, in particular [23, Corollary 3.4], *if G is convex set or a Chebyshev set, then $\mathscr{P}_G$ is IR l.s.c.* Moreover [23, Corollary 3.8], *for a proximinal set G in a strictly convex normed linear space E, $\mathscr{P}_G$ is IR l.s.c. if and only if G is a Chebyshev set*; the necessity part of this result improves the first consequence mentioned after Theorem 5.13 above, where it was assumed that $\mathscr{P}_G$ is l.s.c. Combining this with Theorem 5.9, it follows [23, Corollary 3.9] that *for a boundedly compact set G in a smooth strictly convex Banach space E the following statements are equivalent*: $1°$. *$\mathscr{P}_G$ is IR l.s.c.* $2°$. *$\mathscr{P}_G$ is l.s.c.* $3°$. *G is a Chebyshev set.* $4°$. *G is convex.* B. Brosowski and F. Deutsch have also proved [23, Theorem 3.11] the following sharpening of the global part of Theorem 5.13 above: *If G is a proximinal set in a normed linear space E, such that $\mathscr{P}_G$ is IR l.s.c., then we have* (5.32).

The above results show that if G is a Chebyshev sun (e.g., a convex Chebyshev set) in a normed linear space E, then $\mathscr{P}_G$ (or, one can say, π_G) is OR u.s.c., OR l.s.c. and IR l.s.c. However, for Chebyshev sets these notions do not reduce to the continuity of π_G; actually, we have seen in § 4 many examples of Chebyshev subspaces G with discontinuous metric projection π_G.

Finally, let us mention the following problem, raised by B. Brosowski and F. Deutsch [23].

Problem 5.4. Let G be a proximinal set in a normed linear space E, such that $\mathscr{P}_G$ is OR l.s.c. (i) Is G a sun? (ii) Is $\mathscr{P}_G$ IR l.s.c.? (iii) Is the set $\mathscr{U}_G^{(1)}$ defined by (5.45) dense in E?

For some other properties of $\mathscr{P}_G$ in general metric spaces, see also [168, pp. 377–391] and [148].

(e) We conclude this section with some results and problems on best approximation of operators by elements of nonconvex (hence nonlinear) sets of continuous linear operators.

We have seen in the foregoing that if H is a Hilbert space and $E = L(H, H)$, then the subspaces $\mathscr{C}(H, H)$ of all compact linear operators (§ 2, Theorem 2.13) and $\mathscr{H}(H, H)$ of all Hermitian operators (final part of § 2.2) and the convex cone $\Pi(H, H)$ of all positive operators (final part of § 5.1 (b) above) are proximinal in E. A nonconvex set which is known to be *proximinal* in $E = L(H, H)$ is the set $G = \mathscr{K}_n(H, H)$ *all operators* $u \in E$ *of rank* $\leqq n$ (i.e., such that $\dim u(H) \leqq n$), where $n < \infty$ is fixed (see, for example, [75, Chap. II, § 7, Theorem 7.1], where the distance $\rho(u, G)$ is also computed; the distances $\rho(u, \mathscr{K}_n(H, H)), n = 0,1,2, \cdots$, are called in [75] the "s-numbers" of the operator u and they have various applications in operator theory). However, it is not known whether this proximinality property remains still valid if $E = L(H, H)$ is replaced by $E = L(F_1, F_2)$, where F_1, F_2 are two Banach spaces (see for example, [147]).

R. B. Holmes [84] has raised

Problem 5.5. If H is a Hilbert space, is the set $G = \mathscr{N}(H, H)$ of all normal operators proximinal in $E = L(H, H)$?

Holmes has observed [84] that $G = \mathscr{N}(H, H)$ is a closed nowhere dense cone in $E = L(H, H)$, but it is not convex. Also, Holmes [84] has studied the set $\mathscr{P}_G^{-1}(0)$ for $G = \mathscr{N}(H, H)$ proving, among other results, that no invertible operator and no nonzero compact operator belongs to this set.

V. Pták and M. Fiedler have shown [150, Theorem 1] that *if F is a finite-dimensional Banach space, the set $G = \mathscr{S}(F, F)$ of all singular operators is proximinal in $E = L(F, F)$* (this also follows from the fact $G = \mathscr{S}(F, F)$ is closed and boundedly compact in E) and have computed the distance $\rho(u, G)$, $u \in E$; note that their proof remains valid for any space F such that each invertible operator $u \in L(F, F)$ attains its norm at some point of S_F.

Another problem which has been studied is that of best approximation of (linear or nonlinear) operators by elements of certain nonconvex sets of continuous linear projections. A useful source of proximinality results for this problem is

Theorem 5.7 (and, in particular, Corollary 5.1) above. Y. Ikebe has shown [92] that *if* F_1, F_2 *are two normed linear spaces, then every set* G *in* $E = L(F_1, F_2^*)$, *which is closed in the weak* operator topology on* E (we recall that a net $\{u_d\} \subset E$ is said to be convergent to $u \in E$ in this topology if $(u_d(y))(z) \to (u(y))(z)$ for all $y \in F_1, z \in F_2$), *is proximinal in* E. J. Blatter and E. W. Cheney have observed [14] that this is an immediate consequence of Corollary 5.1 above, since $L(F_1, F_2^*)$ is linearly isometric to the conjugate space $(F_1 \otimes F_2)^*$, where $F_1 \otimes F_2$ is the algebraic (uncompleted) tensor product endowed with the greatest cross-norm γ (see, for example, [157]), by the mapping $u \to f_u$, where $f_u(\sum y_i \otimes z_i) = \sum (u(y_i))(z_i)$, and clearly the image of the weak* operator topology under this linear isometry is nothing else than $\sigma((F_1 \otimes F_2)^*, F_1 \otimes F_2)$. From this it follows immediately ([14], see also [96] and [92]) that *if* F_0 *is a complemented subspace of a normed linear space* F *and if* F_0 *is linearly isometric to a conjugate space* (e.g., if F_0 is a weak* closed subspace of a conjugate space or, in particular reflexive or finite-dimensional), *then the set* $G = P(F, F_0) = \{v \in L(F, F_0) | v(y) = y \, (y \in F_0)\}$ of all continuous linear projections of F onto F_0 *is proximinal.*

Concerning the above subset $G = P(F, F_0)$ of $E = L(F, F_0)$, more attention (than to the problem of proximinality of G) has been given to the minimal and cominimal elements of G, e.g., to the problems of their existence, uniqueness and characterization. We recall that a projection $v_0 \in G = P(F, F_0)$ is called *minimal* if $v_0 \in \mathscr{P}_G(0)$ (or, in other words, if v_0 is an element of minimal norm in G) and *cominimal* if $v_0 \in \mathscr{P}_G(I)$. The numbers $\rho(0, G)$ and $\rho(I, G)$ are also called respectively the *relative projection* and *relative co-projection constant* of F_0 with respect to F and they are called *exact* if $\mathscr{P}_G(0) \neq \emptyset$, respectively $\mathscr{P}_G(I) \neq \emptyset$; clearly, if $G = P(F, F_0)$ is proximinal in $E = L(F, F_0)$, then these constants are exact. In the particular case of a finite-dimensional subspace F_0 of $F = C_{F_1}(Q)$, where Q is a perfect compact space and F_1 a Banach space, the minimal and cominimal projections coincide, since $\|I + v\| = 1 + \|v\|$ for all $v \in P(F, F_0)$ (see § 4.7). E. W. Cheney and K. H. Price [44] have raised the next problem.

Problem 5.6. If F_0 is a proximinal subspace of finite codimension of a normed linear space F, is the projection constant of F_0 with respect to F exact?

The assumption of proximinality of F_0 in Problem 5.6 is essential; indeed, J. R. Isbell and Z. Semadeni have proved [96] that if Q is a compact space containing infinitely many nonisolated points, then $F = C_R(Q)$ contains a nonproximinal closed hyperplane F_0 having the nonexact projection constant 2. J. Blatter and E. W. Cheney have shown [14] that for $F = c_0$ the answers to Problem 5.6 and to the similar problem for the relative co-projection constant are affirmative.

Also known are a few results on uniqueness of minimal projections, for example, that for each subspace F_0 of a Hilbert space H, there is a unique minimal projection $v_0 \in \mathscr{P}_{P(H,F_0)}(0)$. E. W. Cheney, C. R. Hobby, P. D. Morris, F. Schurer and D. E. Wulbert have proved [41] that *for the subspace* F_0 *of* nth *order trigonometric polynomials of the space* $F = C_{2\pi}$ *(of all real* 2π*-periodic functions with the supremum norm), there is a unique minimal projection* $v_0 \in \mathscr{P}_{P(F,F_0)}(0)$, *namely, the Fourier projection.* For other results on this property of the Fourier projection in various spaces, see P. V. Lambert [114], [115].

There also exist some general theorems on characterization of minimal projections (for example, [44, Theorem 3]) and, in particular, of minimal projections onto finite-dimensional subspaces F_0 (for example, [44, Theorem 4]). Some more results are known on characterization of minimal *projections with finite carrier* onto n-dimensional subspaces F_0 in the spaces $F = C_R(Q)$ (Q compact), i.e., of the form $v_0(y) = \sum_{i=1}^{k} y(q_i)y_i$ $(y \in F)$, where $k \geqq n$, $F_0 = [y_1, \cdots, y_k]$ and $q_i \in Q$ for $i = 1, , \cdots, k$ (for example, [39, Theorems 1 and 2]) and, in particular, of minimal *interpolating projections* in these spaces i.e., which are of the above form with $k = n$ and satisfy $v_0(y)(q_i) = y(q_i)$ for all $y \in F = C_R(Q), i = 1, \cdots, n$ (for example, [44, Theorem 12] and [45]). For other results on minimal projections we refer the reader to [42]–[44].

Finally, we mention that there are also a few results on best approximation of nonlinear operators by elements of nonconvex sets of continuous linear projections. For example, using essentially Theorem 5.7 above, Y. Ikebe has proved [92] that *if F_0 is a finite-dimensional Chebyshev subspace of a normed linear space F and $E = l^{\infty}_{F_0}(S_F)$, $G = P(F, F_0)|_{S_F} \subset E$, then $\mathcal{P}_G(\pi_{F_0}) \neq \emptyset$*, i.e., the metric projection onto F_0 has a best approximation by continuous linear projections onto F_0, in the supremum operator norm. For generalizations of this result (e.g., to the set-valued metric projection $\mathcal{P}_{F_0}$) and for other related results see [91], [92].

REFERENCES

[1] N. I. AHIEZER, *Lectures on the Theory of Approximation*. 2nd. ed., Moscow, 1965. (In Russian.)

[2] D. AMIR AND F. DEUTSCH, *Suns, moons and quasi-polyhedra*, J. Approx. Theory, 6 (1972), pp. 176–201.

[3] T. ANDO, *Contractive projections in L_p spaces*, Pacific J. Math., 17 (1966), pp. 391–405.

[4] E. ASPLUND, *Chebyshev sets in Hilbert space*, Trans. Amer. Math. Soc., 144 (1969), pp. 236–240.

[5] D. A. AULT, F. R. DEUTSCH, P. D. MORRIS AND J. E. OLSON, *Interpolating subspaces in approximation theory*, J. Approx. Theory, 3 (1970), pp. 164–182.

[6] R. B. BARRAR AND H. L. LOEB, *On the continuity of the nonlinear Tchebysheff operator*, Pacific J. Math., 32 (1970), pp. 593–601.

[7] C. BESSAGA AND A. PELCZYNSKI, *On extreme points in separable conjugate spaces*, Israel J. Math., 4 (1966), pp. 262–264.

[8] J. H. BIGGS, F. R. DEUTSCH, R. E. HUFF, P. D. MORRIS AND J. E. OLSON, *Interpolating subspaces in l_1-spaces*, to appear.

[9] J. BLATTER, *Zur Stetigkeit von mengenwertigen matrischen Projektionen*, Schriften des Rheinisch-Westf. Inst. für Instrum. Math. Univ. Bonn, Ser. A, No. 16 (1967), pp. 17–38.

[10] ———, *Approximation und Selection*, Habilitationsschrift, Bonn, 1969.

[11] ———, *Approximative Kompaktheit verallgemeinerter rationaler Funktionen*, J. Approx. Theory, 1 (1969), pp. 85–93.

[12] ———, *Weiteste Punkte und nächste Punkte*, Rev. Roumaine Math. Pures Appl., 14 (1969), pp. 615–621.

[13] ———, *Grothendieck spaces in approximation theory*, Mem. Amer. Math. Soc., 120 (1972), pp. 1–121.

[14] J. BLATTER AND E. W. CHENEY, *On the existence of extremal projections*. J. Approx. Theory, 6 (1972), pp. 72–79.

[15] J. BLATTER, P. D. MORRIS AND D. E. WULBERT, *Continuity of the set-valued metric projection*, Math. Ann., 178 (1968), pp. 12–24.

[16] J. BLATTER AND G. L. SEEVER, *The disc algebra is not an existence subspace of its bidual*, to appear.

[17] D. BRAESS, *On topological properties of sets admitting varisolvent functions*, Proc. Amer. Math. Soc., 35 (1972), pp. 453–456.

[18] O. BRANDT, *Geometrische Approximationstheorie in normierten Vektorräumen*, Schriften des Rheinisch-Westf. Inst. für Instrum. Math. Univ. Bonn. Ser. A., No. 18 (1968), pp. 1–36.

[19] B. BROSOWSKI, *Nicht-lineare Tschebyscheff-Approximation*, Bibliogr. Inst. Hochschulskripten Bd. 808/808a, Mannheim, 1968.

[20] ———, *Einige Bemerkungen zum verallgemeinerten Kolmogoroffschen Kriterium*, Funktionalanalytische Methoden der Numerischen Mathematik, ISNM 12, Birkhäuser Verlag, 1969, pp. 25–34.

[21] ———, *Nichtlineare Approximation in normierten Vektorräumen*, Abstract Spaces and Approximation, ISNM 10, Birkhäuser Verlag, 1969, pp. 140–159.

[22] ———, *Fixpunktsätze in der Approximationstheorie*, Mathematica, II (34) (1969), pp. 195–220.

[23] B. BROSOWSKI AND F. DEUTSCH, *Radial continuity of set-valued metric projections*, Ges. Wiss. Datenverarb, MBH Göttingen, 1972 (preprint).

[24] ———, *On some geometrical properties of suns*, Ges. Wiss. Datenverarb. MBH Göttingen, 1972 (preprint).

[25] ———, *Some new continuity concepts for metric projections*, Bull. Amer. Math. Soc., 78 (1972), pp. 974–978.

[26] B. BROSOWSKI, K.-H. HOFFMANN, E. SCHÄFER UND H. WEBER, *Stetigkeitssätze für Metrische Projektionen*, Iterationsverfahren. Numerische Mathematik. Approximationstheorie, ISNM 15, Birkhäuser Verlag, 1970, pp. 11–17.

[27] ———, *Metrische Projektionen auf lineare Teilräume von* $C_0[Q,H]$, Iterationsverfahren. Numerische Mathematik. Approximationstheorie, ISNM 15, Birkhäuser Verlag, 1970, pp. 19–27.

[28] B. BROSOWSKI UND R. WEGMANN, *Charakterisierung bester Approximationen in normierten Vektorräumen*, J. Approx. Theory, 3 (1970), pp. 369–397.

[29] ———, *On the lower semi-continuity of the set-valued metric projection*, Ibid., to appear.

[30] B. BROSOWSKI UND L. WUYTACK, *Charakterisierung und Eindeutigkeit bei der rationalen Approximation mit Nebenbedingungen für die Fehlerfunktion*, Ges. Wiss. Datenverarb. MBH Göttingen, 1972 (preprint).

[31] F. BROWDER, *Multivalued monotone nonlinear mappings and duality mappings in Banach spaces*, Trans. Amer. Math. Soc., 118 (1965), pp. 338–351.

[32] A. L. BROWN, *Best n-dimensional approximation to sets of functions*, Proc. London Math. Soc., 14 (1964), pp. 577–594.

[33] ———, *On continuous selections for metric projections in spaces of continuous functions*, J. Funct. Anal., 8 (1971), pp. 431–449.

[34] ———, *Abstract approximation theory*, Seminar in analysis (1969–1970), Inst. Math. Sci. Madras.

[35] ———, *A discontinuous metric projection in a strictly convex reflexive space*, 1972 (preprint).

[36] ———, *A rotund reflexive space and a subspace of codimension two with a discontinuous metric projection*, to appear.

[37] L. CARLESON AND S. JACOBS, *Best uniform approximation by analytic functions*, to appear.

[38] E. W. CHENEY, *Introduction to Approximation Theory*, McGraw-Hill, New York, 1966.

[39] ———, *Projections with finite carrier*, Proc. Conf. Numerical Methods of Approximation Theory, Oberwolfach, 1971, pp. 19–31.

[40] E. W. CHENEY AND A. A. GOLDSTEIN, *Tchebycheff approximation and related extremal problems*, J. Math. Mech., 14 (1965), pp. 87–98.

[41] E. W. CHENEY, C. R. HOBBY, P. D. MORRIS, F. SCHURER AND D. E. WULBERT, *On the minimal property of the Fourier projection*, Trans. Amer. Math. Soc., 143 (1969), pp. 249–258.

[42] E. W. CHENEY AND P. D. MORRIS, *On the existence and characterization of minimal projections*, Center Numer. Anal. Univ. of Austin, Texas, 1972 (preprint).

[43] ———, *Stability properties of trigonometric interpolating operators*, Center Numer. Anal., Univ. of Austin, Texas, 1972 (preprint).

[44] E. W. CHENEY AND K. H. PRICE, *Minimal projections*, Approximation Theory, A. Talbot, ed., Academic Press, New York, 1970, pp. 261–289.

[45] ———, *Minimal interpolating projections*, Iterationsverfahren, Numerische Mathematik, Approximationstheorie, ISNM 15, Birkhäuser Verlag, 1970, pp. 115–121.

[46] E. W. CHENEY AND D. E. WULBERT, *Existence and unicity of best approximations*, Math. Scand., 24 (1969), pp. 113–140.

[47] F. COMBES, *Sur les faces d'une C*-algèbre*, Bull. Sci. Math., 93 (1969), pp. 37–62.

[48] M. M. DAY, *Normed Linear Spaces*, Springer Verlag, Berlin-Göttingen-Heidelberg, 1972.

[49] F. R. DEUTSCH AND P. H. MASERICK, *Applications of the Hahn-Banach theorem in approximation theory*, SIAM Rev., 9 (1967), pp. 516–530.

[50] F. DEUTSCH AND P. MORRIS, *On simultaneous approximation and interpolation which preserves the norm*, J. Approx. Theory, 2 (1969), pp. 355–373.

[51] F. DEUTSCH, W. POLLUL AND I. SINGER, *On set-valued metric projections, Hahn-Banach extension maps and spherical image maps*, Duke Math J., to appear.

[52] P. DIEROLF, *Korrespondenzen und ihre topologische Eigenschaften*, Überlicke Math. 5, Bibliogr. Inst. Mannheim, to appear.

[53] R. L. DOLGANOV, *Interpolating systems of complex continuous functions*, Sibirsk. Mat. Zh., 12 (1971), pp. 699–706. (In Russian.)

[54] A. I. DUBOVITZKII AND A. A. MILIUTIN, *Extremum problems in the presence of restrictions*, Zh. Vychisl. Mat. i Mat. Fiz., 5 (1965), pp. 395–453. (In Russian.)

[55] N. DUNFORD AND J. SCHWARTZ, *Linear Operators*, Part. I, Interscience, New York, 1958.

[56] M. EDELSTEIN, *Nearest points of sets in uniformly convex Banach spaces*, J. London Math. Soc., 43 (1968), pp. 375–377.

[57] ———, *A note on nearest points*, Quart. J. Math., 21 (1970), pp. 403–406.

[58] M. EDELSTEIN AND A. C. THOMPSON, *Some results on nearest points and support properties of convex sets in* c_0, Pacific J. Math., 40 (1972), pp. 553–560.

[59] N. V. EFIMOV AND S. B. STECHKIN, *Some properties of Chebyshev sets*, Dokl. Akad. Nauk SSSR, 118 (1958), pp. 17–19. (In Russian.)

[60] ———, *Chebyshev sets in Banach spaces*, Abstracts of Short Communications, Internat. Congr. Math. Stockholm, 1962, p. 71. (In Russian.)

[61] G. EWALD, D. G. LARMAN AND C. A. ROGERS, *The directions of the line segments and of the n-dimensional balls on the boundary of a convex body in Euclidean space*, Mathematika, 17 (1970), pp. 1–20.

[62] KY FAN AND I. GLICKSBERG, *Some geometric properties of the spheres in a normed linear space*, Duke Math. J., 52 (1958), pp. 553–568.

[63] KY FAN AND A. HOFFMAN, *Some metric inequalities in the space of matrices*, Proc. Amer. Math. Soc., 6 (1955), pp. 111–116.

[64] C. FOIAŞ AND I. SINGER, *Points of diffusion of linear operators and almost diffuse operators in spaces of continuous functions*, Math Z., 87 (1965), pp. 434–450.

[65] P. FRANCK, *Sur la plus courte distance d'un opérateur aux opérateurs dont le noyau contient un ensemble donné ou est de dimension au moins égale à un nombre donné*, C. R. Acad. Sci. Paris Ser. A-B, 263 (1966), pp. A 388– A 389.

[66] A. L. GARKAVI, *On spaces of linear operators*, Dokl. Akad. Nauk SSSR, 132 (1960), pp. 497–500. (In Russian.)

[67] ———, *The theory of best approximation in normed linear spaces*, Mathematical Analysis 1967, Moscow, 1969, pp. 75–132. (In Russian.)

[68] ———, *The problem of Helly and best approximation in the space of continuous functions*, Izv. Akad. Nauk SSSR, 31 (1967), pp. 641–656. (In Russian.)

[69] ———, *Compacta admitting Chebyshev systems of measures*, Mat. Sb., 74 (116) (1967), pp. 209–217. (In Russian.)

[70] ———, *Characterization of Chebyshev subspaces of finite codimension in* L_1, Mat. Zametki, 7 (1970), pp. 155–163. (In Russian.)

[71] ———, *The problem of Helly and best approximation of summable functions*, Mat. Sb., 84 (126), 2 (1971), pp. 196–217. (In Russian.)

[72] G. GODINI, *Best approximation in normed linear spaces by elements of convex cones*, Studii şi cercet. mat., 21 (1969), pp. 931–936. (In Romanian.)

[73] ———, *Semi-continuity properties of set-valued metric projections*, Rev. Roumaine Math. Pures Appl., to appear.

[74] ———, *Characterizations of proximinal linear subspaces in normed linear spaces*, Ibid., to appear.

[75] I. C. GOHBERG AND M. G. KREIN, *Introduction to the Theory of Linear Non-Selfadjoint Operators*, Moscow, 1965. (In Russian.) English translation: Amer. Math. Soc., Providence, R.I., 1969.

[76] A. HAAR, *Die Minkowskische Geometrie und die Annäherung an stetige Funktionen*, Math. Ann., 78 (1918), pp. 294–311.

[77] H. HAHN, *Reelle Funktionen*, 1 Teil, Akademische Verlagsgesellschaft, Leipzig, 1932.

[78] P. R. HALMOS, *Positive approximants of operators*, Indiana Univ. Math. J., 21 (1972), pp. 951–960.

[79] S. IA. HAVINSON, *On approximation by elements of convex sets*, Dokl. Akad. Nauk SSSR, 172 (1967), pp. 294–297. (In Russian.)

[80] R. B. HOLMES, *Smoothness indices for convex functions and the unique Hahn-Banach extension problem*, Math. Z., 119 (1971), pp. 95–110.

[81] ———, *On the continuity of best approximation operators*, Proc. Symp. Infinite Dimensional Topology, Ann. of Math. Studies 69, Princeton Univ. Press, Princeton, 1972, pp. 137–157.

[82] ———, *A course on optimization and best approximation*, Lecture Notes 257, Springer Verlag, Berlin-Heidelberg-New York, 1972.

[83] ———, *Smoothness of certain metric projections on Hilbert space*, to appear.

[84] ———, *Best approximation by normal operators*, to appear.

[85] R. B. HOLMES AND B. R. KRIPKE, *Approximation of bounded functions by continuous functions*, Bull. Amer. Math. Soc., 71 (1965), pp. 896–897.

[86] ———, *Interposition and approximation*, Pacific J. Math., 24 (1968), pp. 103–110.

[87] ———, *Smoothness of approximation*, Michigan Math. J., 15 (1968), pp. 225–248.

[88] ———, *Best approximation by compact operators*, Indiana Univ. Math. J., 21 (1971), pp. 255–263.

[89] R. B. HOLMES AND J. LAMBERT, *A geometrical approach to property (SAIN)*, J. Approx. Theory, to appear.

[90] R. B. HOLMES AND J. D. WARD, *An approximative property of spaces of continuous functions*, to appear.

[91] Y. IKEBE, *A study of operators in approximation theory*, Dissertation, Univ. Texas, 1966.

[92] ———, *Generalizations of the Alaoglu theorem with applications to approximation theory*, Proc. Japan Acad., 44 (1968), I, pp. 439–441; II, pp. 442–445.

[93] ———, *A characterization of Haar subspaces in* $C[a, b]$, Proc. Japan, Acad., 44 (1968), pp. 219–220.

[94] ———, *A characterization of best Tchebycheff approximations in function spaces*, Ibid., 44 (1968), pp. 485–488.

[95] A. D. IOFFE AND V. M. TIHOMIROV, *Duality of convex functions and extremal problems*, Uspehi Mat. Nauk, 23, 6(144) (1968), pp. 51–116. (In Russian.)

[96] J. R. ISBELL AND Z. SEMADENI, *Projection constants and spaces of continuous functions*, Trans. Amer. Math. Soc., 107 (1963), pp. 38–48.

[97] R. C. JAMES, *Characterizations of reflexivity*, Studia Math., 23 (1964), pp. 205–216.

[98] ———, *A counterexample for a sup theorem*, Israel J. Math., 9 (1971), pp. 511–512.

[99] ———, *Reflexivity and the sup of linear functionals*, Ibid., 13 (1972), pp. 289–300.

[100] M. I. KADEC, *Topological equivalence of uniformly convex spaces*, Uspehi Mat. Nauk., 10, 4(66) (1955), pp. 137–141. (In Russian.)

[101] J. P. KAHANE, *Projection métrique de* $L^1(T)$ *sur des sous-espaces fermés invariants par translation*, Linear Operators and Approximation, ISNM 20, Birkhäuser Verlag, Basel-Stuttgart, 1972, pp. 302–309.

[102] V. KLEE, *The support property of a convex set in a linear normed space*, Duke Math. J., 15 (1948), pp. 767–772.

[103] ———, *A characterization of convex sets*, Amer. Math. Monthly, 56 (1949), pp. 247–249.

[104] ———, *Research problem no.* 5, Bull, Amer. Math. Soc., 63 (1957), p. 419.

[105] ———, *Convexity of Chebyshev sets*, Math. Ann., 142 (1961), pp. 292–304.

[106] ———, *Remarks on nearest points in normed linear spaces*, Proc. Coll. Convexity (Copenhagen, 1965), Univ. of Copenhagen, 1966, pp. 168–176.

[107] ———, *Can the boundary of a d-dimensional convex body contain segments in all directions?* Amer. Math. Monthly, 76 (1969), pp. 408–410.

[108] A. N. KOLMOGOROV, *A remark on the polynomials of P. L. Chebyshev deviating the least from a given function*, Uspehi Mat. Nauk, 3, 1(23) (1948), pp. 216–221. (In Russian.)

[109] C. A. KOTTMAN AND BOR-LUH LIN, *On the weak continuity of metric projections*, Michigan Math. J., 17 (1970), pp. 401–404.

[110] B. R. KRIPKE AND T. J. RIVLIN, *Approximation in the metric of $L^1(X, \mu)$*, Trans. Amer. Math. Soc., 119 (1965), pp. 101–122.

[111] B. R. KRIPKE AND R. T. ROCKAFELLAR, *A necessary condition for the existence of best approximations*, J. Math. Mech., 13 (1964), pp. 1037–1038.

[112] J. KRUSKAL, *Two convex counterexamples: a discontinuous envelope function and a non-differentiable nearest point mapping*, Proc. Amer. Math. Soc., 23 (1969), pp. 697–703.

[113] J. M. LAMBERT, *The weak sequential continuity of the metric projection in L_p spaces*, Dissertation, Purdue Univ., 1970.

[114] P. V. LAMBERT, *Minimum norm property of the Fourier projection in spaces of continuous functions*, Bull. Soc. Math. Belg., 21 (1969), pp. 359–369.

[115] ———, *On the minimum property of the Fourier projection in L^1 spaces*, Ibid., 21 (1969), pp. 370–391.

[116] P. J. LAURENT, *Théorèmes de caractérisation en approximation convexe*, Mathematica, 10 (33) (1968), pp. 95–111.

[117] ———, *Conditions nécessaires pour une meilleure approximation non linéaire dans un espace normé*, C. R. Acad. Sci. Paris Sér. A-B, 269 (1969), pp. A 245–A 248.

[118] ———, *Approximation et optimisation*, Hermann, Paris, 1972.

[119] P. J. LAURENT AND PHAM-DINH-TUAN, *Global approximation of a compact set by elements of a convex set in a normed space*, Numer. Math., 15 (1970), pp. 137–150.

[120] A. J. LAZAR, *Spaces of affine continuous functions on simplexes*, Trans. Amer. Math. Soc., 134 (1968), pp. 503–525.

[121] A. J. LAZAR, D. E. WULBERT AND P. D. MORRIS, *Continuous selections for metric projections*, J. Functional Analysis, 3 (1969), pp. 193–216.

[122] J. LINDENSTRAUSS, *Extension of compact operators*, Mem. Amer. Math. Soc., 48 (1964).

[123] ———, *On nonlinear projections in Banach spaces*, Michigan Math. J., 11 (1964), pp. 263–287.

[124] ———, *On nonseparable reflexive Banach spaces*, Bull. Amer. Math. Soc., 72 (1966), pp. 967–970.

[125] J. LINDENSTRAUSS AND L. TZAFRIRI, *On the complemented subspace problem*, Israel J. Math., 9 (1971), pp. 263–269.

[126] G. G. LORENTZ, *Approximation of Functions*, Holt, Rinehart and Winston, New York, 1966.

[127] G. MEINARDUS, *Approximation of Functions, Theory and Numerical Methods*, Springer Verlag, Berlin-Heidelberg-New York, 1967.

[128] E. A. MICHAEL, *Continuous selections, I*, Ann. Math., 63 (1956), pp. 361–382.

[129] J. J. MOREAU, *Fonctionnelles convexes*, Séminaire sur les équations aux dérivées partielles, Collège de France, Paris, 1966–67.

[130] ———, *Distance à un convexe d'un espace normé et caractérisation des points proximaux*, Séminaire d'analyse unilatérale, Fac. Sci. Montpellier, 2 (1969), exposé no. 6.

[131] P. D. MORRIS, *Metric projections onto subspaces of finite codimension*, Duke Math. J., 35 (1968), pp. 799–808.

[132] ———, *Approximation theory*, Lecture notes, Univ. of Austin, Texas, 1971.

[133] W. NITZSCHE, *Über die mengenwertige Abbildung $\mathscr{P}_G$ und drei Schemata zur Approximationstheorie*, Hahn-Meitner-Inst. für Kernforschung Berlin GmbH, Report BM-43, 1971 (preprint).

[134] C. OLECH, *Approximation of set-valued functions by continuous functions*, Colloq. Math., 19 (1968), pp. 285–293.

[135] E. V. OSHMAN, *Continuity of metric projections and some geometric properties of the unit sphere in a Banach space*, Dokl. Akad. Nauk SSSR, 185 (1969), pp. 34–36. (In Russian.)

[136] ———, *On continuity of metric projection in a Banach space*, Mat. Sb., 80 (122), 2(10) (1969), pp. 181–194. (In Russian.)

[137] ———, *Chebyshev sets, continuity of metric projection and some geometric properties of the unit sphere in a Banach space*, Izv. Vyssh. Uchebn. Zaved. Matematika, 4(83) (1969), pp. 38–46. (In Russian.)

[138] ———, *Chebyshev sets and continuity of metric projection*, Ibid., 9(100) (1970), pp. 78–82. (In Russian.)

[139] ———, *On continuity of metric projection onto some classes of subspaces in a Banach space*, Dokl. Akad. Nauk SSSR, 195 (1970), pp. 555–557. (In Russian.)

[140] ———, *On a criterion of continuity of metric projection in a Banach space*, Mat. Zametki, 10 (1971), pp. 459–468. (In Russian.)

[141] J. M. OVERDECK, *On the nonexistence of complex Haar systems*, Bull. Amer. Math. Soc., 77 (1971), pp. 737–740.

[142] A. PELCZYNSKI, *A generalization of Stone's theorem on approximation*, Bull. Acad. Pol. Sci., 5 (1957), pp. 105–107.

[143] ———, *Linear extensions, linear averagings and their applications to linear topological classification of spaces of continuous functions*, Dissert. Math. (Rozprawy Mat.) 58, Warszawa, 1968.

[144] R. R. PHELPS, *Convex sets and nearest points*, Proc. Amer. Math. Soc., 8 (1957), pp. 790–797.

[145] ———, *Uniqueness of Hahn-Banach extensions and unique best approximation*, Trans. Amer. Math. Soc., 95 (1960), pp. 238–255.

[146] ———, *Chebyshev subspaces of finite dimension in L_1*, Proc. Amer. Math. Soc., 17 (1966), pp. 646–652.

[147] A. PIETSCH, *Einige neue Klassen von kompakten linearen Abbildungen*, Rev. Roumaine Math. Pures Appl., 8 (1963), pp. 427–447.

[148] W. POLLUL, *Topologien auf Mengen von Teilmengen und Stetigkeit von mengenwertigen metrischen Projektionen*, Diplomarbeit, Bonn, 1967.

[149] ———, *Reflexivität und Existenz-Teilräume in der linearen Approximationstheorie*, Dissertation, Bonn (1971), Schriften der Ges. für Math. und Datenverarbeitung, Bonn, 53 (1972), pp. 1–21.

[150] V. PTÁK ET M. FIEDLER, *Sur la meilleure approximation des transformations linéaires par des transformations de rang prescrit*, C. R. Acad. Sci. Paris, 254 (1962). pp. 3805–3807.

[151] J. R. RICE, *The Approximation of Functions. Vol. I: Linear Theory. Vol. II: Non-linear and Multivariate Theory*, Addison-Wesley, Reading, Mass.-London-Don Mills, Ont., 1964 and 1969.

[152] ———, *Nonlinear Approximation*, Approximation of Functions, H. L. Garabedian, ed., Elsevier, Amsterdam-London-New York, 1965, pp. 111–133.

[153] ———, *Nonlinear approximation. II. Curvature in Minkowski geometry and local uniqueness*, Trans. Amer. Math. Soc., 128 (1967), pp. 437–459.

[154] W. W. ROGOSINSKI, *Continuous linear functionals on subspaces of $\mathscr{L}^p$ and $\mathscr{C}$*, Proc. London Math. Soc., 6, 22 (1956), pp. 175–190.

[155] CH. ROUMIEU, *Sur quelques problèmes d'approximation*, Sémin. math. fac. sci. Montpellier, 1966.

[156] G. SH. RUBINSTEIN, *On an external problem in a linear normed space*, Sibirsk. Mat. Zh., 6 (1965), pp. 711–714. (In Russian.)

[157] R. SCHATTEN, *A Theory of Cross-Spaces*, Ann. of Math. Studies no. 26, Princeton Univ. Press, Princeton, 1950.

[158] Z. SEMADENI, *Simultaneous extensions and projections in spaces of continuous functions*, Lecture notes, Aarhus University, Denmark, 1965.

[159] ———, *Banach Spaces of Continuous Functions*, Vol. I, Monografje Matematyczne 55, Warszawa, 1971.

[160] B. SENDOV, *Certain questions in the theory of approximations of functions and sets in the Hausdorff metric*, Uspehi Mat. Nauk, 24, 5(149) (1969), pp. 141–178. (In Russian.)

[161] ———, *Approximation relative to Hausdorff distance*, Approximation Theory, A. Talbot, ed., Academic Press, London-New York, 1970.

[162] I. SINGER, *Properties of the surface of the unit cell and applications to the solution of the problem of uniqueness of the polynomial of best approximation in arbitrary Banach spaces*, Studii şi cercet. mat., 7 (1956), pp. 95–145. (In Romanian.)

[163] ———, *Caractérisation des éléments de meilleure approximation dans un espace de Banach quelconque*, Acta Sci. Math., 17 (1956), pp. 181–189.

[164] ———, *On best approximation of continuous functions*, Math. Ann., 140 (1960), pp. 165–168.

[165] ———, *On best approximation of continuous functions. II*, Rev. Math. Pures et Appl., 6 (1961), pp. 507–511.

[166] ———, *Some remarks on approximative compactness*, Rev. Roum. Math. Pures et Appl., 9 (1964), pp. 167–177.

[167] ———, *On the extension of continuous linear functionals and best approximation in normed linear spaces*, Math. Ann., 159 (1965), pp. 344–355.

[168] ———, *Best approximation in normed linear spaces by elements of linear subspaces*, Publ. House Acad. Soc. Rep. Romania, Bucharest, 1967. (In Romanian.) English translation: Publ. House Acad. Soc. Rep. Romania, Bucharest and Springer Verlag, Grundlehren Math. Wiss. 171, Berlin-Heidelberg-New York, 1970.

[169] ———, *Some open problems on best approximation in normed linear spaces*, Séminaire Choquet, 6e année. Université de Paris, 1966/67, exposé no. 12.

[170] ———, *On metric projections onto linear subspaces of normed linear spaces*, Proc. Conf. Projections and Related Topics, Clemson, 1967. Preliminary Edition, January, 1968.

[171] ———, *Remark on a paper of Y. Ikebe*, Proc. Amer. Math. Soc., 21 (1969), pp. 24–26.

[172] ———, *On normed linear spaces which are proximinal in every superspace*, J. Approx. Theory, to appear.

[173] ———, *Bases in Banach spaces*, Vol. I, Grundlehren Math. Wiss. 154, Springer Verlag, Berlin-Heidelberg-New York, 1970.

[174] ———, *On set-valued metric projections*, Linear Operators and Approximation, ISNM 20, Birkhäuser Verlag, Basel-Stuttgart, 1972, pp. 217–233.

[175] ———, *Best approximation in normed linear spaces*, CIME Lectures, Erice, Italy, June–July 1971 (to appear in Constructive Functional Analysis, Edizioni Cremonese, Roma).

[176] ———, *On best approximation in normed linear spaces by elements of subspaces of finite codimension*, Rev. Roumaine Math. Pures Appl., 17 (1972), pp. 1245–1256.

[177] ———, *Generalizations of methods of best approximation to convex optimization in locally convex spaces. I: Extension of continuous linear functionals and characterizations of solutions of continuous convex programs*, to appear.

[178] S. B. STECHKIN, *On approximation of abstract functions*, Rev. Math. Pures Appl., 1 (1956), pp. 79–84. (In Russian.)

[179] ———, *Approximative properties of sets in normed linear spaces*, Ibid., 8 (1963), pp. 5–18. (In Russian.)

[180] IU. A. SHASHKIN, *Interpolating families of functions and embeddings of sets into euclidean and projective spaces*, Dokl. Akad. Nauk SSSR, 174 (1967), pp. 1030–1032. (In Russian.)

[181] K. TATARKIEWICZ, *Une théorie généralisée de la meilleure approximation*, Ann. Univ. Mariae Curie-Sklodowska, 6 (1952), pp. 31–46.

[182] L. P. VLASOV, *Approximatively convex sets in Banach spaces*, Dokl. Akad. Nauk SSSR, 163 (1965), pp. 18–21. (In Russian.)

[183] L. P. VLASOV, *On Chebyshev sets*, Ibid., 173 (1967), pp. 491–494. (In Russian.)

[184] ———, *Approximatively convex sets in uniformly smooth spaces*, Mat. Zametki, 1 (1967), pp. 443–449. (In Russian.)

[185] ———, *On Chebyshev and approximatively convex sets*, Ibid., 2 (1967), pp. 191–200. (In Russian.)

[186] ———, *Chebyshev sets and some generalizations of them*, Ibid., 3 (1968), pp. 59–69. (In Russian.)

[187] ———, *Approximative properties of sets in Banach spaces*, Ibid., 7 (1970), pp. 593–604. (In Russian.)

[188] ———, *Almost convex and Chebyshev sets*, Ibid., 8 (1970), pp. 545–550. (In Russian.)

[189] ———, *Some theorems on Chebyshev sets*, Ibid., 11 (1972), pp. 135–144. (In Russian.)

[190] R. WEGMANN, *Zur Stetigkeit der mengenwertigen metrischen Projektion in endlichdimensionalen Räumen*, Manuscr. Math., 7 (1972), pp. 373–386.

[191] ———, *Some properties of the peak-set-mapping*, to appear.

[192] D. E. WULBERT, *Continuity of metric projections*, Trans. Amer. Math. Soc., 143 (1968), pp. 335–343.

[193] D. E. WULBERT, *Convergence of operators and Korovkin's theorem*, J. Approx. Theory, 1 (1968), pp. 8–18.

[194] ———, *Differential theory for non-linear approximation* (preprint).

[195] ———, *Uniqueness and differential characterization of approximations from manifolds of functions*, Amer. J. Math., 18 (1971), pp. 350–366.

[196] ———, *Nonlinear approximation with tangential characterization*, to appear.

[197] E. ZARANTONELLO, *Projections on convex sets in Hilbert space and spectral theory*, Contributions to Non-Linear Functional Analysis, E. Zarantonello, ed., Academic Press, New York-London, 1971, pp. 237–424.

[198] S. I. ZUKHOVITSKII, *On minimal extensions of linear functionals in the space of continuous functions*, Izv. Akad. Nauk SSSR, 21 (1957), pp. 409–422. (In Russian.)

[199] S. I. ZUKHOVITSKII AND G. I. ESKIN, *Some theorems on best approximation by unbounded operator-functions*, Ibid., 24 (1960), pp. 93–102. (In Russian.)

Complements added in proof.

1. *To page* 22: G. W. Henderson and B. L. Ummel have proved [208] that *a locally connected compact space Q admits a complex Chebyshev system $x_1, \cdots, x_n$ with $n \geqq 2$ if and only if Q is homeomorphic to a subset of the plane.* This substantiates a conjecture of J. M. Overdeck [141] and generalizes a result of I. J. Schoenberg and C. T. Yang (see [168]) in which Q was assumed to be finite polyhedral.

2. *To page* 28: Proposition 3.2 has been generalized to regular subsets (see p. 73 for their definition) of $C_H(Q)$, where Q is a compact metric space and H is a Hilbert space, by K.-H. Hoffmann [209]. Recently W. Pollul has proved [220, Theorem 1] that *if a normed linear space E has the property that for every $f \in E^*$ with $\|f\| = 1$ and every pair of distinct $x, y \in \mathcal{M}_f$ there exists an element $z \in E$ with $f(z) = 0$ and $x + z$, $x + z + 2(y - x)/\|y - x\| \in \mathcal{M}_f$, then every linear subspace G of E with property* (3.13) *is a semi-Chebyshev subspace.* The spaces $C_R(Q)$ (Q compact), $L^1_R(T, v)$ ((T, v) a positive measure space) and $C^1_R([a, b], v)$ (v the Lebesgue measure) belong to the above class [220]. Also, W. Pollul has proved [220, corollary of Theorem 2] that *for a normed linear space E the following statements are equivalent*: 1°. *Every hyperplane G of E with property* (3.13) *is a semi-Chebyshev subspace.* 2°. *If $f \in E^*$, $\|f\| = 1$ and $\mathcal{M}_f$ contains at least two elements, then $\mathcal{M}_f$ also contains two elements of distance* 2.

3. *To page* 35: D. Sarason has proved (see [206, Lemma 5]) that $\rho(x, H^\infty) = \rho(x, A)$ *for all* $x \in C(\mathrm{Fr}\, D)$. Recently T. Gamelin, J. Garnett, L. Rubel and A. Shields have proved (see [206, Corollary 7]) that *A is not a proximinal subspace of $C(\mathrm{Fr}\, D, v)$*.

4. *To pages* 36–37: A generalization of (3.20) to set-valued metric projections has been given by H. W. McLaughlin and K. B. Somers [218]. For related results see also [217].

5. *To page* 39: R. Holmes, B. Scranton and J. Ward have proved [210, Theorem 1] that *if* $E = L(H, H)$ *and* $G = \mathscr{C}(H, H)$, *where H is a Hilbert space, then* $\dim \mathscr{P}_G(x) = \infty$ *for all* $x \in E \setminus G$ (and thus, in particular, G is a very non-Chebyshev subspace of E); nevertheless [210, Theorem 2], $\mathscr{P}_G^{-1}(0)$ is nowhere dense in E. See also [211], [212] for related results.

6. *To page* 44: J. M. Lambert has proved [215] that there exists a (reflexive) strictly convex isomorph E of l^2 (namely, the dual of a space constructed by V. Klee [212]), with the following properties: (i) E^* is not Fréchet smooth; (ii) for every closed linear (hence Chebyshev) subspace G of E, π_G is continuous; (iii) for $G_0 = [e_{2n}] \subset E$ (where e_n is the nth unit vector in E), $\pi_{G_0}^{-1}(0)$ is not boundedly compact and $t|_{G_0^\perp}$ is not continuous. Consequently, in the necessity parts of Theorems 4.5 and 4.6 the assumption that codim $G < \infty$ cannot be removed.

To page 62: Some results on weak semi-continuity of set-valued metric projections have been obtained recently by G. Godini [207].

8. *To pages* 63–65: Another sufficient condition for the existence of continuous selections, with application to continuous selections for Hahn–Banach extension maps (hence also for metric projections, by Corollary 4.7), has been given by J. Lindenstrauss [216]. Recently H. Fakhoury has proved [205], Corollary 5 that *if* $G = M$ *is an "M-ideal" of a Banach space E*, i.e., a closed linear subspace such that there exists a linear projection u of E^* onto $G^\perp$ satisfying $\|f\| = \|u(f)\| + \|f - u(f)\|$ for all $f \in E^*$, *then* $\mathscr{P}_G$ *admits a continuous selection*; for example, any two-sided ideal G in a C^*-algebra E (in particular, $G = \mathscr{C}(H, H)$ in $E = L(H, H)$, where H is a Hilbert space) is an M-ideal of E. Also, H. Fakhoury has proved [205, Theorem 9] that if Q, S are two compact spaces and $E = C(Q)$, $G = \psi^0(C(S))$, where $\psi^0 : C(S) \to C(Q)$ is the "conjugate" mapping defined in § 2, formula (2.17), then $\mathscr{P}_G$ admits a continuous selection. Some other recent results on continuous selections, Borel measurable selections, quasi-additive selections and homogeneous selections for metric projections have been obtained by G. Nürnberger [219].

9. *To page* 65: For real scalars Theorem 5.1 has also been proved by F. Deutsch [202].

10. *To page* 66: Some results on best approximation in normed linear spaces by elements of finite-dimensional closed convex cones and, in particular, a characterization of the elements of best approximation by the cone (5.6), when a certain sufficient condition for uniqueness is satisfied, have been obtained by R. J. Duffin and L. A. Karlovitz [203].

11. *To page* 67: E. Asplund has proved [200] that *the metric projection onto any closed subset of a finite-dimensional Euclidean space is almost everywhere differentiable.* This solves a problem of J. Kruskal [112].

12. *To page* 77: Problem 5.1 has been solved affirmatively by M. Edelstein [57, first paragraph], namely, if $E = l_R^2 \times R$ with $\|\{x, \lambda\}\| = \max(\|x\|, |\lambda|)$ and

$G = \{\{e_n, 1/n\} | n = 1, 2, \cdots\}$, where e_n is the nth unit vector in l_R^2, then $\mathscr{D}(\pi_G)$ is not dense in E (e.g., $\{0, -2\}$ has a neighborhood which is disjoint from $\mathscr{D}(\pi_G)$).

13. *To page* 78: Problem 5.2 has been solved in the negative by C. B. Dunham [204], who has constructed an example of a Chebyshev set which is not a sun in $C([0, 1])$. The idea of Dunham has been extended by B. Brosowski, F. Deutsch, J. Lambert and P. D. Morris [201], who have shown, among other results, that if T is a σ-compact, locally compact space which contains a compact nowhere dense G_δ-set (e.g., if T is an infinite compact metric space), then $C_0(T)$ contains a one-parameter family of Chebyshev sets which are not suns.

14. *To page* 83: Problem 5.5 has been solved in the negative by D. D. Rogers [221], who has shown that the following sets G_i of operators are non-proximinal in $E = L(H, H)$, where H is a Hilbert space: (i) $G_1 = \mathscr{N}(H, H)$, the set of all normal operators; (ii) G_2 = the set of all unitary operators; (iii) G_3 = the set of all (continuous linear) projections; (iv) G_4 = the set of all isometries; (v) G_5 = the set of all diagonalizable operators; (vi) G_6 = the set of all compact normal operators.

15. We note also the appearance, while the present book was in press, of the book of H. Kiesewetter [213], which treats some topics in the theory of best approximation in normed linear spaces.

Finally, we wish to express our thanks to all colleagues who attended the National Science Foundation Regional Conference on The Theory of Best Approximation and Functional Analysis at Kent State University, 11–15 June, where this monograph was presented, for their stimulating interest and valuable comments.

REFERENCES TO THE "COMPLEMENTS ADDED IN PROOF"

[200] E. Asplund, *Differentiability of the metric projection in finite-dimensional Euclidean space*, Proc. Amer. Math. Soc., 38 (1973), pp. 218–219.

[201] B. Brosowski, F. Deutsch, J. M. Lambert and P. D. Morris, *Chebyshev sets which are not suns*, to appear.

[202] F. Deutsch, Thesis, Brown University, 1965.

[203] R. J. Duffin and L. A. Karlovitz, *Formulation of linear programs in analysis. I: Approximation theory*, SIAM J. Appl. Math., 16 (1968), pp. 662–675.

[204] C. B. Dunham, *Chebyshev sets in* $C[0, 1]$ *which are not suns*, Canad. Math. Bull., to appear.

[205] H. Fakhoury, *Projections de meilleure approximation continues dans certains espaces de Banach*, C. R. Acad. Sci. Paris Ser. A, 276 (1973), pp. A 45–A 48.

[206] T. Gamelin, J. Garnett, L. Rubel and A. Shields, in preparation.

[207] G. Godini, to appear.

[208] G. W. Henderson and B. L. Ummel, *The nonexistence of complex Haar systems on nonplanar locally connected spaces*, Proc. Amer. Math. Soc., 39 (1973), pp. 640–641.

[209] K.-H. Hoffmann, *Über ein Eindeutigkeitskriterium bei der Tschebyscheff-Approximation mit regulären Funktionensystemen*, Funktionalanalytische Methoden der numerischen Mathematik, Oberwolfach 1967, pp. 71–79. Birkhäuser Verlag, Basel, 1969.

[210] R. B. Holmes, B. E. Scranton and J. D. Ward, *Best approximation by compact operators. II*, preprint.

[211] ———, *The metric complement of the compact operators*, preprint.

[212] ———, *Uniqueness of commuting compact approximations*, preprint.

[213] H. Kiesewetter, *Vorlesungen über lineare Approximation*, VEB Deutscher Verlag der Wissenschaften, Berlin, 1973.

[214] V. KLEE, *Two renorming constructions related to a question of Anselone*, Studia Math., 33 (1969), pp. 213–242.
[215] J. M. LAMBERT, *Continuous metric projections*, Proc. Amer. Math. Soc., to appear.
[216] J. LINDENSTRAUSS, *A selection theorem*, Israel J. Math., 2 (1964), pp. 201–204.
[217] H. W. McLAUGHLIN AND K. B. SOMERS, *Another characterization of Haar subspaces*, to appear.
[218] ———, *Characterizations of generalized strong unicity in approximation theory*, to appear.
[219] G. NÜRNBERGER, *Schnitte für die metrische Projektion*, Diplomarbeit, Erlangen, 1973.
[220] W. POLLUL, *Über ein Eindeutigkeitskriterium für beste Approximationen*, to appear.
[221] D. D. ROGERS, *Proximinal sets of operators*, preprint.